Inquiry-Based Sciei in Grades 6–12

This new book shows middle and high school science teachers how to use evidence-based inquiry to help students achieve deeper conceptual understanding. Drawing on a wealth of research, authors Patrick Brown and James Concannon demonstrate how direct, hands-on experience in the science classroom can enable your students to become more self-reliant learners. They also provide a plethora of model lessons aligned with the Next Generation Science Standards (NGSS) and offer advice on how to create your own lesson plans and activities to satisfy the demands of your curriculum. With the resources in this book, you and your students will be able to ditch the textbook and embark upon an exciting and rewarding journey to scientific discovery.

Patrick Brown, PhD, is the STEM coordinator for the Fort Zumwalt School District in O'Fallon, Missouri.

James Concannon, PhD, is a science teacher educator at Westminster College in Fulton, Missouri.

Also Available by Routledge Eye On Education

www.routledge.com/eyeoneducation

STEM by Design: Strategies and Activities for Grades 4–8
Anne Jolly

The STEM Coaching Handbook: Working with Teachers to Improve Instruction
Terry Talley

Creating Scientists: Teaching and Assessing Science Practice for the NGSS
Christopher Moore

Writing Science Right: Strategies for Teaching Scientific and Technical Writing
Sue Neuen and Elizabeth Tebeaux

Write, Think, Learn: Tapping the Power of Daily Student Writing Across the Content Areas
Mary K. Tedrow

Inquiry-Based Science Activities in Grades 6–12

Meeting the NGSS

Patrick Brown and James Concannon

Routledge
Taylor & Francis Group

NEW YORK AND LONDON

First published 2018
by Routledge
711 Third Avenue, New York, NY 10017

and by Routledge
2 Park Square, Milton Park, Abingdon, Oxon, OX14 4RN

Routledge is an imprint of the Taylor & Francis Group, an informa business

© 2018 Taylor & Francis

Library of Congress Cataloging-in-Publication Data
A catalog record for this title has been requested

ISBN: 978-0-8153-8336-9 (hbk)
ISBN: 978-0-8153-8337-6 (pbk)
ISBN: 978-1-3510-6458-3 (ebk)

Typeset in Palatino and Myriad Pro
by Apex CoVantage, LLC

Dedication

To family and friends who gave their unending support and
encouragement and especially to

Finn, Lua, and my wife Cathy – PB

Sophia, Jackson, Brett, and my wife Joy – JC

The lesson plans comprising this book were originally published in *Science Activities: Classroom Projects and Curriculum Ideas*, a peer-reviewed journal that provides teachers and educators with the best classroom-tested projects, experiments, and curriculum ideas that promote scientific inquiry through active learning experiences.

For more information about *Science Activities*, including tips on how to prepare and submit a manuscript for publication, visit: http://tandfonline.com/action/journalInformation?show=aimsScope&journalCode=vsca20.

Contents

List of Figures

List of Tables

Meet the Authors

Patrick L. Brown, PhD, is the STEM Coordinator for the Fort Zumwalt School district in St. Charles, Missouri. A former middle and high school teacher, Dr. Brown has taught science methods courses for prospective elementary, middle, and high school teachers and presented extensively on many of his teaching and research projects in science classrooms. His science teaching ideas have appeared in *Science and Children*, *Science Scope*, *The Science Teacher*, and *Science Activities*. His research in science education has been published in *Science Education*, the *Journal of Science Teacher Education*, and the *International Journal of Science Education*.

Dr. James Concannon is an Associate Professor of Education at Westminster College where he teaches elementary methods of teaching science, middle and secondary school methods of teaching science, secondary school teaching, foundations of education, and supervises clinical teaching experiences. Dr. Concannon's professional accomplishments include various teaching awards, research awards, and publishing in a variety of journals including *International Journal of Science Education*, *Journal of STEM Education*, *School Science and Mathematics*, *Journal of Science Education and Technology*, *Science Scope*, and *Science Activities: Classroom Projects and Curriculum Ideas*. Prior to earning his PhD., Dr. Concannon taught high school chemistry, physics, earth science, zoology, and human physiology. In the fall of 2018, Dr. Concannon is joining the education faculty at William Woods University (Fulton, Missouri) as an Associate Professor of STEM Education.

Acknowledgements

We are very thankful to many science educators, teachers, and students who over the years have reshaped our understanding of effective teaching and learning. We also have some specific thank yous.

I would like to express my sincere appreciation to the Fort Zumwalt School District. I was supported by administrators, coordinators, teachers, and students too numerous to thank individually here, but whose influence is evident in the narrative in this book. – PB

My Westminster College colleagues have contributed in numerous positive ways to the creation of this book. I am grateful for all their contributions. I am also looking forward to transitioning to William Woods University and to continued and new collaborations. – JC

Introduction

Promoting evidence-based inquiry in your classroom is a lot like building a bridge. Whether the bridge is an arch, beam, suspension, or cabled-stayed, the signature feature—ability to support objects from point A to B—is of upmost importance. Bridge builders are accountable for ensuring bridges do not "fail under loading." They need to consider the strength, weight, and cost of materials, resonance due to stress, and fluctuations because of temperature and wind changes. Like bridge building, we want our science classes to be supportive and provide connective learning experiences for students learning science. Evidence-driven inquiry is built on the premise that teachers must scaffold classroom activities so that students use their immediate experiences with data and evidence (Point A) to arrive at a scientifically accurate claim about the phenomenon under study (Point B).

The analogy that evidence-driven inquiry is like a bridge cannot be highlighted enough. A remarkable thing occurs for students when teachers use evidence-driven inquiry. Students direct experiences and the claims they construct serve as the basis for their science understanding. In addition, students' evidence-based claims act as an easy entry point for introducing key science vocabulary, concepts, and supporting ideas at a meaningful time. From a cognitive standpoint, students' conceptual understanding of science is linked directly to their experiences with data and evidence. By sequencing lessons where students are provided opportunities to collect data and use this as evidence to support conclusions, teachers open opportunities for students to experience essential science practices. Additional student-centered elaborations after making evidence-based claims should be directly connected to students' experiences. This elaboration process is critical because it allows students to develop much richer and sophisticated ways of thinking about the world based on their direct experiences, requiring students to transfer knowledge to new and unique situations. Thus, evidence-driven inquiry empowers students with deep conceptual understanding and cultivates the skills they need to more self-reliant learners.

How to Use This Book

This book grew out of our work with K-12 students as classroom science teachers. Through purposeful and intentional lesson planning, we learned important aspects of instructional design that have broad implications for learning. We also learned that many of the tried-and-true, research-based science teaching practices naturally translate to reform advocated by *A Framework for K-12 Science Education (The Framework)* and the *Next Generation Science Standards* (National Research Council [NRC] 2011; NGSS Lead States 2013). In Chapter 1, we provide a short but informative description of three research-based pathways to highly effective science teaching. Chapter 2 offers an overview of the NGSS. Chapters 1 and 2 are closely connected and the pathways described in Chapter 1 easily allow for the translation into the NGSS. Thus, the combination of Chapters 1 and 2 show teachers how to apply "tried-and-true" research-based practices in science education to put the NGSS into practice. The bulk of the book provides lessons that use one of more of the three pathways to effective science teaching. Educators are encouraged to use the ideas in Chapter 2 (e.g., overview of the NGSS) to unpack the Model Lessons by identifying components of the NGSS that occur during teaching and learning. These ideas presented here were tested in science teacher professional development and through our work with prospective teachers in science methods courses. Below we share many of our lessons learned along the way.

Examples: Teacher Professional Development

Many teachers use demonstrations to teach science content only to discover that they do not have the lasting impact on students' understanding of the content. We have found success by supporting teachers to restructure demonstrations in instructional sequences that promote students making evidence-based claims. In one lesson presented here, students use the PSOE model to understand weather and climate. We used a PSOE sequence of instruction that included the following phases: Predict, Share, Observe, and Explain. In the PSOE phases, students first commit to an outcome based on their prior knowledge. Then students explore before they explain the outcome. This instructional sequence taps into students' prior knowledge. In addition, by having students observe before explaining, the PSOE sequence is an easy pathway to helping drive students' evidence-based claims. During the weather and climate demonstration, students developed a long-lasting understanding of the transfer of energy and "convection" currents. Many teachers we've worked with comment that simple shifts in the arrangement of activities, such as doing a demonstration before explaining new content, is a way to promote deep conceptual understanding. Teachers like the PSOE model because students construct knowledge firsthand based on their experiences with data. The model lesson easily addresses the three dimensions of the Frameworks and NGSS.

Examples: Beginning Teachers

One of the biggest challenges beginning teachers face is seamlessly transitioning from one lesson to the next. The model lesson "How thin is foil? Applying density to find the thickness of aluminum foil" is one example of how to structure multiple days' worth of instruction to support learning. In this model lesson, an inquiry-based approach teaches fundamental concepts associated with understanding density. Beginning teachers should take note that throughout this model lesson one of the most powerful learning experiences occurs when students collect data to generate scientifically accurate claims. Once students have constructed knowledge and articulated it in their own words, the teacher can connect new terms, key ideas, and other related concepts to students' foundational knowledge. In this regard, beginning teachers benefit from the structure of the model lesson because it is an explicit example of how to construct lessons in accordance with how students learn science best (Bransford, Brown, and Cocking 2000; Donovan and Bransford, 2005).

Conclusion

All of the model lessons were tested with students (and many teachers). The model lessons can easily be implemented in classroom settings, through professional development experiences, or in science methods courses. Teachers at all stages of the professional continuum benefit from reflecting on the activities and effectiveness of the lessons individually and in groups. Teachers have commented that the model lessons are a starting point for reflecting on learning and that they provide them with a plethora of ideas for their classroom. For many, implementing and reflecting on the model lessons has helped develop their professional knowledge of learners and learning, instructional activities, and understanding of content. In addition, numerous teachers successfully adapt these ideas to design their own, unique evidence-driven inquiries for students.

References

Bransford, J., A. Brown, and R. Cocking. 2000. *How people learn: Brain, mind, experience, and school*. Washington, DC: National Academy Press.

Donovan, M.S., and J.D. Bransford., eds. 2005. *How students learn: Science in the classroom*. National Research Council Committee on How People Learn: A targeted report for teachers. www.nap.edu/catalog/11102/how-students-learn-science-in-the-classroom.

NGSS Lead States. 2013. *Next Generation Science Standards: For states, by states*. Washington, DC: National Academies Press. Retrieved from www.nextgenscience.org/next-generation-sciencestandards.

National Research Council. 2012. *A framework for K–12 science education: Practices, crosscutting concepts, and core ideas*. Washington, DC: National Academies Press.

1

What Are the Features
of Evidence-Driven Inquiry?

Kids love to explore the world and explain how nature works. They form ideas about the causes for the changing of the seasons, create theories about the phases of the moon, and try to describe why they have some physical characteristics like their parents. These are just a few of the many science areas that students have ideas about based on their experiences. Students at a very early age think logically about their environment and look for patterns and relationships to form explanations for science. Regardless of the accuracy of their ideas, students' immediate experiences are the basis for how they know and understand the world they live in.

While students' scientific understandings can be a great starting point for instruction, they can also act as barrier for gaining knowledge. Research in the cognitive sciences and science education demonstrate the implications of students' prior knowledge, and particularly their misconceptions, on learning (Bransford, Brown, and Cocking 2000; Donovan and Bransford, 2005). Prior knowledge is an important consideration in teaching, and students' incoming ideas, including misconceptions, must be addressed in order for them to gain more accurate and complete science understanding. In fact, many resources are dedicated to identifying typical misconceptions in many different science disciplines (see Driver, Squires, and Wood-Robinson 1994) and several books offer engaging ways to elicit students' science views (see Keeley and Tugel 2009; Keeley, Eberle, and Dorsey 2008; Keeley, Eberle, and Farrin 2005; Keeley, Eberle, and Tugel 2007). The reason why prior knowledge is so important in teaching relates back to the early 1980s and conceptual change research.

This continued line of research clearly shows that the most powerful and influential instructional sequences require a purposeful interaction between students' incorrect or partially incomplete ideas and direct experiences to develop more plausible, intelligible, and fruitful explanations (Posner et al. 1982). For students to accommodate new information, they must first become dissatisfied with their current conceptions, and this is achieved by teachers providing opportunities for students to collect data and scientific evidence that cannot be explained when students rely upon their incomplete understandings. Learning facts is not enough to improve students' understanding of science. To understand science, students need opportunities to view new ideas in broader contexts of meaning.

From a conceptual change perspective, instruction should start with assessing students' incoming ideas. If a teacher's entry point into a lesson does not begin with students' prior knowledge, conceptual misunderstandings may arise whereby students simply assimilate new information into their existing inaccurate foundation of knowledge (National Research Council 1997; Posner et al. 1982). By knowing students' prior knowledge and experiences, teachers can choose the best types of experience to create dissatisfaction for incorrect ideas so that students can begin to construct conceptions that are more accurate. The best experiences are ones whereby students are provided with evidence-based experiences. Because students have great capacity to reason at very sophisticated levels from teaching approaches that productively scaffold their developing content knowledge and science reasoning skills, how teachers provide "hands-on" science for students requires deeper exploration.

Evidence-Driven Inquiry

If the ultimate goal is for students to derive understanding from experiences, then we must carefully consider our professional practices. While hands-on learning can naturally be engaging for students, the experiences must be carefully weaved into the flow of instruction to produce the desired outcomes. What are the desired incomes? An important finding from America's Lab report is that many students view science as a "false dichotomy," meaning that students think that the hands-on, "doing" part of science is separate from content (Singer, Hilton, and Schweingruber 2006). As a result, the desired outcomes are for students to discard incorrect ideas, accept the most accurate scientific explanations, and for students to learn the nature by which these science explanations are generated. Evidence-driven inquiry allows teachers to meet these goals by first providing students with immediate experiences to form accurate understandings; and second, by connecting student's claims to scientifically accepted explanations. Connections are established when teachers purposefully link evidence from explorations to evidence-based explanations. Explanations can be further supported by lecture, readings, and discussions. In sum, evidence-driven inquiry requires a special combination of students' evidence-based experiences, students' scientific claims, and the teacher connecting students' claims to our current understandings of science phenomena. Three interrelated educational ideas, referred to here as "pathways," are catalysts for promoting evidence-driven inquiry.

Pathway 1: Sequencing Science Instruction

Science instruction should be sequenced where students explore prior to teachers introducing science terminology, ideas, or concepts. The learning cycle is an approach where students explore prior to any introduction of science terminology or formal explanation. The learning cycle includes three sequential phases (1) *exploration*, (2) *invention (term introduction)*, (3) *discovery (concept application)* (Karplus and Their 1967). The learning cycle was initially created to align to Piagetian stages of assimilation, disequilibrium, and accommodation (Treagust and Tsui 2014). When employing the learning cycle, students have experiences with data (exploration) that is then used by them to construct accurate evidence-based claims (student portion of invention phase). Students' evidence-based claims are the foundation for their understanding and used to introduce key science terms, concepts, and supporting ideas (teacher portion of invention phase). Once students have constructed knowledge and have authoritative explanations (e.g., teacher lectures, textbook readings, discussions, etc. that occur during the teacher portion of the invention phase), they are given the opportunity to practice and test out new knowledge in new and different situations (discovery phase). Thus, the learning cycle places primacy on students' exploratory experiences from which they construct some aspect of science knowledge at a conceptual level. Student-constructed knowledge is used as an "anchor" for learning related topics. Studies have compared the learning cycle to variations that sequence the teachers' explanation and the beginning of instruction and use investigations to verify provided ideas. As result of a learning cycle process, students use science vocabulary accurately and can explain valid and reliable ways to generate ideas about their everyday world. This line of scholarship shows the learning cycle sequence to be more effective at promoting science achievement, motivation, and encouraging scientific reasoning than any other variation (Abraham 1992; Abraham and Renner 1986; Gerber, Cavollo, and Marek 2001; Purser and Renner 1983; Renner, Abraham, and Birnie 1988). Since the initial invention of the learning cycle, other models have been created that retain the exploration before explanation sequence such as the POE (Predict,

Observe, and Explain) and the 5Es (Engagement, Exploration, Explanation, Elaboration, and Evaluation) (Bybee 1997). By sequencing lessons using an Explore before Explain instruction sequence, teachers can create a student-centered learning environment where students ask questions, plan and conduct investigations, gather data, and make evidence-based explanations.

Pathway 2: Wedding Science Content with Practices
Evidence-based inquiry is an approach where students seamlessly learn science content and the nature by which scientific knowledge is produced in valid and reliable ways. The nature by which scientific knowledge is produced has long been an important learning standard. The practices of how science knowledge is developed has been termed "inquiry" and is described by the National Research Council (NRC 2000) as consisting of five essential features:

1. The learner engages in scientifically oriented questions.
2. The learner gives priority to evidence.
3. The learner formulates explanations based on evidence.
4. The learner connects explanations to scientific knowledge.
5. The learner communicates and justifies explanations.

The five essential features of inquiry describe the interconnected processes that scientists use to describe the natural world. In teaching, inquiry is a multifaceted activity that requires students to use critical thinking skills and make connections between evidence and scientific knowledge.

Three perspectives of inquiry include science teachers' implementation of essential features of inquiry into instruction, students learning science in an inquiry-based classroom, and students learning about scientific inquiry (Lederman and Niess 2000). Students are motivated in an inquiry classroom because they perceive instruction as relevant, transferable, and useful in future problem-solving situations (Anderson 1997). Though inquiry is much more complex than hands-on science activities (Crawford 2000), studies have found that hands-on is a critical component and improves students' attitudes and student achievement (Freedman 1997). Studies also show that inquiry-based instruction helps close the gender, ethnic, and socioeconomic status gap with respect to science achievement (Von Secker and Lissitz 1999).

Arguably, one can perceive that scientific inquiry is at the core of science literacy (Songer, Lee, and McDonald 2003) whereby students are engaged in scientifically oriented questions, explore ideas, develop procedures, and make evidence-based decisions in relation to current scientific theory (Newman et al. 2004). Crawford (2000) explains that significant components of an effective inquiry-based classrooms are: "instruction situated in authentic problems; (teachers and students) focus on grappling with data, collaboration of students and teacher, connections with society, teacher modeling behaviors of scientists, and development of student ownership" (p. 933). Teachers have to take on multiple roles to achieve this, a few being a motivator, guide, modeler, learner, and collaborator (Crawford 2000).

Inquiry in the classroom spans a continuum ranging from structured to guided, and from guided to open/independent (Windschitl 2003). A structured approach would be more appropriate for incorporating inquiry with a classroom of younger or inexperienced learners. Structured inquiry requires more facilitation and guidance from the teacher whereby the problem, methods of investigation, and the answers are provided by the teacher (Schwab 1962; Herron 1971). Though structured inquiry requires much more teacher facilitation, the role of the teacher is not to provide a verification-type experience. Verification occurs when science teachers explain the expectations of an activity or a laboratory. The sure-sign of verification is when a science teacher says, "This is what you should expect if you do the lab right."

At the other end of the spectrum, open inquiry requires less direction from the teacher and more decisions to be made by students. In open inquiry, students are more responsible for developing and carrying out their own independent scientific investigations in a process that is much more intellectually challenging for learners (Windschitl 2003). Walking into a classroom where learners are engaged in open inquiry, one would see multiple elements of authentic scientific research in the classroom. From students developing research questions, creating procedures and models, collecting data, using data as evidence for assertions, evaluating the reliability and validity of the data collection process, to considering various aspects of error and ways to limit error in the procedure are just a few possible accounts of how students would be engaged in undertaking science.

Pathway 3: Using Science Phenomena as the Foundation for Learning

A third component in promoting evidence-driven inquiry is honing in on the phenomena that students explore. The idea behind phenomena-based teaching is to focus students' experiential learning on science experiences that lead to wonderment about the natural world. Beneficial explorations invoke curiosity and promote investigations that produce empirical data or qualitative observation and accurate evidence-based claims. In addition, meaningful science phenomena are complex entities that are best understood through multiple learning experiences (e.g., explorations, lectures, readings, discussions, etc.) and allow other key, associated ideas to be easily connected to students' experiences. Science phenomena are not factoids and are important because they contextualize all science learning during a topic of study.

Phenomena-based teaching can be associated with "project-based" and "problem-based" learning. According to Krajcik and his colleagues, the following characteristics are beneficial ways to motivate kids in phenomena-based activities:

1. Driving question that engages students in investigating an authentic problem or situation
2. Collaborative learning around the phenomena under study
3. Technology appropriate to the problem
4. Products that represent what students have learned (Krajcik et al. 1994).

Research in these areas indicates that using the authentic and meaningful contexts that phenomena-based teaching provides engages students in developing deep conceptual understanding (Brown and Abell 2007).

What many find is that the three pathways naturally intersect, and that the design principles associated with one directly relates to all pathways. Each of the three "pathways" have been major themes in *Ready, Set, Science! Putting Research to Work in K-8 Science Classrooms* (Michaels, Shouse, and Schweingruber 2008). Although students' background experiences vary from child to child, using the pathways as core approaches to science teaching is a highly effective approach to develop all kids' knowledge. Before we jump right into the model lessons, let's make one more connection to the current vision of science education reform.

References

Abraham, M.R. 1992. Instructional strategies designed to teach science. In F. Lawrenz, K. Cochran, J. Krajcik, and P. Simpson, eds. *Research matters. . . To the science teacher* (pp. 41–50.). Manhattan, KS: NARST Monograph #5.

Abraham, M.R., and J.W. Renner. 1986. The sequence of learning cycle activities in high school chemistry. *Journal of Research in Science Teaching, 23,* 21–43.

Anderson, O.R. 1997. A neurocognitive perspective on current learning theory and science instructional strategies. *Science Education, 81*(1), 67–89.

Bransford, J., A. Brown, and R. Cocking. 2000. *How people learn: Brain, mind, experience, and school.* Washington, DC: National Academy Press.

Brown, P., and S. Abell. 2007. Examining the learning cycle. *Science and Children, 44*(5), 58–59.

Bybee, R.W. 1997. *Achieving scientific literacy: From purposes to practices.* Portsmouth, NH: Heinemann Educational Books, Inc.

Crawford, B.A. 2000. Embracing the essentials of inquiry: New roles for science teachers. *Journal of Research in Science Teaching, 37,* 916–937.

Donovan, M.S., and J.D. Bransford, eds. 2005. *How students learn: Science in the classroom.* National Research Council Committee on How People Learn: A targeted report for teachers. www.nap.edu/catalog/11102/how-students-learn-science-in-the-classroom.

Driver, R., A. Squires, P. Rushworth, and V. Wood-Robinson. 1994. *Making sense of secondary science.* London: Routledge.

Freedman, M. 1997. Relationship among laboratory instruction, attitude toward science, and achievement in science knowledge. *Journal of Research in Science Teaching, 34*(4), 343–357.

Gerber, B.L., A.M.L. Cavallo, and E.A. Marek. 2001. Relationship among informal learning environments, teaching procedures, and scientific reasoning abilities. *International Journal of Science Education, 23,* 535–549.

Herron, M.D. 1971. The nature of scientific inquiry. *School Review, 79*(2), 171–212.

Karplus, R., and H.D. Their. 1967. *A new look at elementary school science.* Chicago: Rand McNally.

Krajcik, J., P. Blumenfeld, R. Marx, and E. Solloway. 1994. A collaborative model for helping middle grade teachers learn project-based instruction. *The Elementary Science Journal, 94*(5), 483–498.

Keeley, P., and J. Tugel. 2009. *Uncovering student ideas in science: 25 new formative assessment probes,* vol. 4. Arlington, VA: NTSA Press.

Keeley, P., F. Eberle, and J. Tugel. 2007. *Understanding student ideas in science: 25 more formative assessment* probes, vol. 2. Arlington, VA: NSTA Press.

Keeley, P., F. Eberle, and L. Farrin. 2005. *Understanding student ideas in science: 25 formative assessment probes,* vol. 1. Arlington, VA: NSTA Press.

Keeley, P., F. Eberle, J. Tugel., and C. Dorsey. 2008. *Uncovering student ideas in science: Another 25 formative assessment probes,* vol. 3. Arlington, VA: NTSA Press.

Lederman, N. G., and M.L. Niess. 2000. Problem solving and solving problems: Inquiry about inquiry. *School Science and Mathematics, 100*(3) 113–116.

Michaels, S., A.W. Shouse, and H.A. Schweingruber. 2008. *Ready, set, science! Putting research to work in K-8 science classrooms.* Board on Science Education, Center for Education, Division of Behavioral and Social Science and Education. Washington, DC: The National Academies Press. www.nap.edu/catalog/11882/ready-set-science-putting-research-to-work-in-k-8#toc.

Newman, W.J., S.K. Abell, P.D. Hubbard, J. McDonald, J. Otaala, and M. Martini. 2004. Dilemmas of teaching inquiry in elementary science methods. *Journal of Science Teacher Education, 15,* 257–279.

Posner, G. J., K.A. Strike, P.W. Hewson, and W.A. Gertzog. 1982. Accommodation of a scientific conception: Toward a theory of conceptual change. *Science Education, 66,* 211–227.

Purser, R. K., and J.W. Renner. 1983. Results of two tenth-grade biology teaching procedures. *Science Education, 67,* 85–98.

National Research Council (NRC). 2000. *Science teaching reconsidered: A handbook.* Washington, DC: The National Academics Press.

Renner, J. W., M.R. Abraham, and H.H. Birnie. 1988. The necessity of each phase of the learning cycle in teaching high school physics. *Journal of Research in Science Teaching*, *25*, 39–58.

Schwab, J.J. 1962. The teaching of science as inquiry. In J.J. Schwab and P.F. Brandwein, eds. *The teaching of science* (pp. 3–103). Cambridge, MA: Harvard University Press.

Singer, S.R., M.L. Hilton, and H.A. Schweingruber, eds. 2006. *America's lab report: Investigations in high school science*. Washington, DC: National Academies Press.

Songer, N.B., H.S. Lee, and S. McDonald. 2003. Research towards an expanded understanding of inquiry science beyond an idealized standard. *Science Education*, *87*, 490–516.

Treagust, D. F., and C.Y. Tsui. 2014. General instructional methods and strategies. In S.K. Abell and N.G. Lederman, eds. *Handbook of research on science education*, vol. 2 (pp. 373–391). Mahwah, N.J.: Lawrence Erlbaum Associates.

Von Secker, C., and R. Lissitz. 1999. Estimating the impact of instructional practices on student achievement in science. *Journal of Research in Science Teaching*, *36*(10), 1110–1126.

Windschitl, M. 2003. Inquiry projects in science teacher education: What can investigative experiences reveal about teacher thinking and eventual classroom practice? *Science Education*, *87*, 112–143.

2

Next Generation Science Standards (NGSS)

As many veteran teachers know, with time comes educational change. Release of the Next Generation Science Standards (NGSS) raised many questions for teachers. Some of the first questions that came to our teachers' minds were, *"How is the NGSS going to change my teaching? How do I convert the NGSS into practice?"* While the new standards reflect advancement in the science education community, we have found that using one or more of the three pathways previously described in daily classroom activities allows for the seamless translation of the NGSS into practice. Using the pathways opens opportunities for teachers to address the three interconnected dimensions of NGSS: Science and Engineering Practices, Crosscutting Concepts, and Disciplinary Core Ideas.

Science and Engineering Practices, Crosscutting Concepts, and Disciplinary Core Ideas for K-12 Curriculum

Practices

The eight essential science and engineering practices embody the vision for what students should know and be able to do to understand the world in which they live (summarized in Bybee 2012). The revision, in part, reflects the difficulty teachers had in embracing "inquiry-based" teaching due to misconceptions between what "inquiry" means in teaching and in everyday vernacular (Bybee 2012). Similar to the essential features of classroom inquiry, the eight science and engineering practices (SEP) can vary from being teacher- to student-directed for a range of implementation strategies. The practices are overlapping, meaning that asking questions leads to designing and conducting experiments, and then to making sense of observations derived from experiences with data and evidence.

Crosscutting Concepts

The *crosscutting concepts* (CCC) have explanatory power across the sciences and help bridge the SEPs with the Disciplinary Core Ideas. The CCCs provide an organizational framework for helping students connect knowledge from different disciplines (summarized in Duschl 2012). For example, students can use observed "patterns" and "cause-and-effect" to bridge their experiences with data to arrive at evidence-based claims. In this example, pattern recognition in data and seeking to understand the mechanisms behind cause and effect are useful logical thinking tools that help students understand the natural world in much the same way that scientists carry out their work. Thus, the CCCs are embedded in the SEPs and content under study.

Disciplinary Core Ideas

The final dimension is the one most familiar to teachers and is the content. The Disciplinary Core Ideas (DCIs) include key ideas in major science fields (e.g., earth and space science, physical science, life science). Each of the DCIs include component ideas that offer more specific information related to content standards.

Connecting the Three Pathways to the NGSS

It is helpful to expand our view of effective teaching when translating the NGSS into classroom instruction. The "pathways" to evidence-based inquiry and the anatomy of the standards are closely connected (teachers may find it helpful to go back and forth between this section and Table 2.1). For instance, consider how the explore before explain instructional sequence connects to the NGSS. Many explorations ask students to investigate scientific questions (alignment with SEPs 1 and 3). Students' explorations produce data that requires analysis (alignment with SEP 4). According to the data collected, students think about patterns and trends, and cause and effect relationships (all CCC) that can help students make evidence-based claims (alignment with

Table 2.1 The Three-Dimensional Frameworks

Science and Engineering Practices	Crosscutting concepts
1. Asking questions (for science) and defining problems (for engineering) 2. Developing and using models 3. Planning and carrying out investigations 4. Analyzing and interpreting data 5. Using mathematics and computational thinking 6. Constructing explanations (for science) and designing solutions (for engineering) 7. Engaging in argument from evidence 8. Obtaining, evaluating, and communicating information	1. Patterns 2. Cause and effect: Mechanism and explanation 3. Scale, proportion, and quantity 4. Systems and system models 5. Energy and matter: Flows, cycles, and conservation 6. Structure and function 7. Stability and change

DCIs

Physical Science

- MS-PS1 Matter and its Interactions
- MS-PS2 Motion and Stability: Forces and Interactions
- MS-PS3 Energy
- MS-PS4 Waves and their Applications in Technologies for Information Transfer

Life Science

- MS-LS1 From Molecules to Organisms: Structures and Processes
- MS-LS2 Ecosystems: Interactions, Energy, and Dynamics
- MS-LS3 Heredity: Inheritance and Variation of Traits
- MS-LS4 Biological Evolution: Unity and Diversity

Earth and Space Science

- MS-ESS1 Earth's Place in the Universe
- MS-ESS2 Earth's Systems
- MS-ESS3 Earth and Human Activity

Engineering, Technology, and Applications in Science

- MS-ETS1 Engineering Design

SEP 6). Students' evidence-based claims should be about science content that is either directly related to or an elaboration of the curriculum (alignment with DCIs). These are just a few of the easy ways the explore before explain sequence translates to the NGSS.

A similar transition to NGSS occurs when teachers use an "inquiry" pathway in their classroom. Consider essential features of inquiry 1 and 2 which explain that learners should "engage in scientifically oriented questions" and "give priority to evidence." By engaging students in scientifically oriented questions, teachers are covering many SEPS (alignment with 1, 3, 4, and 5). In addition, similar to the explore before explain sequence, CCC are embedded in the "inquiry" activity as a necessary part of developing understanding. When teachers ask students to "connect explanations to scientific knowledge" and "communicate and justify explanation", they are engaging student in the SEPs 6, 7, and 8. In sum, inquiry directly translates to NGSS dimensions in many ways.

Finally, phenomena-based approaches also directly relate to the NGSS. A key aspect of phenomena-based teaching is investigating science in order to form deep conceptual understanding. If students' investigations are hands-on and produce data, they would be engaging in many potential SEPs and CCCs. In addition, the "theme" or phenomenon of interest should relate to a teacher's curriculum, and therefore be related to DCIs. Teachers who hone in on meaningful and relevant topics that require hands-on exploration open up inclusion of all three dimensions of the NGSS.

Meaningful Professional Development

While some teachers and professional developers (and school districts alike) have success in developing curriculum and lessons directly from the standards, others are not so fortunate. Few activities are more powerful than reflection on practice. We have noticed that educators learn best when they have opportunities to collaborate, have active learning experiences, and can focus on student learning. The model lessons that follow serve as great reflective tools for teachers to develop a deeper understanding of the NGSS in practice. While NGSS documents provide an extensive overview of the infrastructure of the integration of SEPs, CCCs, and DCIs, the model lessons provide examples of how to put the vision of the standards into practice.

We have success in teaching educators how to unpack standards by using the model lessons that illustrate the three dimensions of the framework. Teachers, professional developers, science educators, and curriculum teams can carry out effective professional learning by implementing the model lessons with students and then reflecting on practice.

One of the most beneficial activities educators can do is reflect on teaching and create NGSS "connection" tables (see Table 2.2 below as an example). NGSS connection tables allow teachers to align activities used in the model lessons with their corresponding dimensions of the NGSS. Teachers have learned quite a bit about their practice from connecting the model lessons to the NGSS. Some teachers who use the pathways receive validation that their existing practices align with the NGSS. For others, the "pathways" provide a more-research-based structure for creating lessons and curriculum. Regardless, having teachers connect standards to model lessons is the type of activity that gives them purpose and intention for understanding three-dimensional learning. Learning is similar for students and teachers. Just like students benefiting from learning science that is relevant to their lives, the model lessons contextualize the NGSS for instructional practice.

Example 1: Translating a Model Lesson to the NGSS

In the model lesson "Teaching Bernoulli's principle through demos," we use classic demonstrations during our forces in fluids activities to help students go back and forth between their observations of phenomena and what occurs at the microscopic level using what we have termed "molecular talk." This model lesson helps students to overcome their misconceptions about airflow and air pressure and their thoughts that blowing air on an object will always cause it to move away. Eliciting students' incoming ideas and misconceptions, and providing evidence-based explorations, helps promote the conceptual change process, making this lesson a highly beneficial learning experience for students. There are numerous chances for teacher formative assessments so they can scaffold learning activities to build students' knowledge and many chances for students to reflect on their developing understanding. This model lesson lends itself to developing more coherent curriculums that connect ideas from physical sciences that are necessary for understanding earth science.

Table 2.2 Unwrapping the Standards for the Bernoulli's Model Lesson

MS-PS2. Motion and Stability: Forces and Interactions
www.nextgenscience.org/dci-arrangement/ms-ps2-motion-and-stability-forces-and-interactions

Performance Expectation	Connections to Classroom Activity
Plan an investigation to provide evidence that the change in an object's motion depends on the sum of the forces on the object and the mass of the object.	• Students use this model lesson to gain deeper understanding of the PE in relation to the forces created when fluids, such as air, with different pressures, encounter each other.
Science and Engineering Practices	**Connections to Classroom Activity**
Asking questions and defining problems	• Students predict what will happen when the teacher uses a straw to blow into the paper tent and between two soda cans resting on straws (Activities 1 and 2). Students' experiences in Activities 1 and 2 lead to new questions explored in Activities 6 and 7.

(Continued)

Table 2.2 *(Continued)*

Developing and using models	• Students were asked to predict and explain what happens in the paper tent and soda cans demonstration using "molecular talk" (Activities 1 and 2). Students develop understanding that is more elaborate by viewing a computer simulation that explains the air pressure when comparing molecules moving in random directions versus in the same direction. Students can use the simulation to make predictions about what happens to an object that is between air moving in the same direction and air molecules randomly encountering each other.
Analyzing and interpreting data	• Students use date from their observations to explain what happens when moving air interacts with objects (paper tent and soda cans on straws) (Activities 1 and 2). Students use data observed in Activities 6 and 7 to verify their experiences in Activities 1 and 2.
Planning and carrying out investigations	• Students carry out investigates to learn about the effects of moving air and air pressure on objects (Activities 1 and 2). Students test out ideas about airflow in new and different situations (Activities 6 and 7).
Constructing explanations and design solutions	• Students formulate evidenced-based claim statements about air pressure moving from "high" to "low." Students' explanations were verified through Activities 1, 2, 6, and 7.
Obtaining, evaluating and communicating information	• Students share their predictions and rudimentary theories about the results of air flow on objects. Students communicate their developing understanding by constructing evidenced-based claims.
Disciplinary Core Idea	**Connections to Classroom Activity**
PS2.A: Forces and Motion The motion of an object is determined by the sum of the forces acting on it; if the total force on the object is not zero, its motion will change. All positions of objects and the directions of forces and motions must be described in an arbitrarily chosen reference frame and arbitrarily chosen units of size. In order to share information with other people, these choices must also be shared.	• Students learn that blowing creates a "stream of air molecules" that are all moving in the same direction. Students explain that in a "stream of air" molecules are all moving in the same directions and the collisions among molecules are fewer compared to air molecules in stationary air in the system under study. In other words, as a flowing fluid gains speeds, the internal pressure in the fluid decreases in relation to the surrounding air.

(Continued)

Table 2.2 *(Continued)*

Crosscutting Concepts	Connections to Classroom Activity
Patterns	• Students work together to look for patterns between their observations of airflow and the influence on objects.
Cause and effect	• Students think in terms of the cause and effect relationships between airflow and the direction an object moves.
Bundling PEs	**Connections to Classroom Activity**
MS-ESS 2–5 Collect data to provide evidence for how the motions and complex interactions of air masses results in changes in weather conditions	• The PSOE demonstrations is a great way to engage students in the forces at work when moving air encounters air systems where the molecules are moving randomly. Students' experiences during the PSOE demonstrations directly related to understanding the interactions that occur when different air masses collide. This PSOE model lesson is an easy way to develop students' understanding of weather from a deeper, conceptual level.

(NGSS Lead States 2013)

Example 2: Translating a High School Model Lesson to the NGSS

In the model lesson "Windmills by Design: Purposeful Curriculum Design to Meet Next Generation Science Standards in a 9–12 Physics Classroom," we use engineering design challenges to help students use their science observations and create a windmill that can better meet the design challenges of the task. We have found that using the windmill lesson in the 5E instructional sequence easily allows for translation into the NGSS. Having students explore and conduct iterative testing on windmills before explaining energy transformation allows them to engage in many SEPS (6 in total) and 1 CCC in a meaningful and relevant science context. Thus, the 5E lesson effectively integrates the three dimensions of the NGSS (SEPS, CCCs, and DCIS). Finally, although the windmill lesson is just one step towards helping students become more proficient at the PE listed in Table 2.3, it connects with many other Performance Expectations related to understanding energy (HS-PS3 Energy). We have shared how the model lesson can be bundled with another PE (e.g., bundling HS-PS3–3 with HS-PS3–1) to form a more coherent curriculum.

Table 2.3 Unwrapping the Standards for the Windmill Model Lesson

HS-PS3-Energy
www.nextgenscience.org/dci-arrangement/hs-ps3-energy

Performance Expectation	Connections to Classroom Activity
HS-PS3–3: Design, build, and refine a device that works within given constraints to convert one form of energy into another form of energy.*	• Throughout the lesson, students design, construct, and evaluate the performance of windmill blades with the goal of producing the most power to learn about the engineering design process and how energy can be transformed

Science and Engineering Practices	Connections to Classroom Activity
Asking questions (for science) and defining problems (for engineering)	• Students are engaged with a problem-based challenge to design windmill blades that will produce the greatest amount of power.
Developing and using models	• Students sketch blade designs and envisage windmill models based on their research about windmill physics. Students must explain the forces involved with windmill design and be able to make predictions about how changes in one part of the windmill may influence other parts of the windmill.
Planning and carrying out investigations	• After researching the physics behind windmill design, students carry out an investigation to test windmills. Students decide the number of windmill blades, how to cut and sand windmill blades, and how to attach the windmill blades to produce the greatest amount of power.
Analyzing and interpreting data	• Based on a time measurement, and assuming constant acceleration, students determine the amount of force the windmill exerts on the weight and the amount of power the windmill produced at varying fan speeds and varying the distances the fan is from the windmill.
Constructing explanations (for science) and designing solutions (for engineering)	• After the first exploration, students reconsider how to make a better design and retest. The combination of testing, rebuilding, and retesting allows students to make evidenced-based claims about optimal windmill design and factors influencing efficiency.
Obtaining, evaluating, and communicating information	• Throughout the lesson, students compare designs, collaborate in groups, and communicate their findings, reflecting upon lessons learned with respect to windmill design.

(Continued)

Table 2.3 (Continued)

Disciplinary Core Idea	Connections to Classroom Activity
PS3.A: Definitions of Energy At the macroscopic scale, energy manifests itself in multiple ways, such as in motion, sound, light, and thermal energy. **PS3.D: Energy in Chemical Processes** Although energy cannot be destroyed, it can be converted to less useful forms — for example, to thermal energy in the surrounding environment	• Students see first-hand how wind energy is transferred to rotational kinetic energy of blades and heat as a result of rotational friction, to linear kinetic energy of the string, and to potential energy of the mass after it is lifted.

Crosscutting Concepts	Connections to Classroom Activity
Constructing explanations and designing solutions	• Students learn that an important attribute of many science disciplines is that through iterative testing they can form deeper scientific explanations and more comprehensive design solutions. Students' knowledge of the engineering design process and evidenced-based experiences leads to science knowledge that is directly connected to their experiences with windmills. For example, students modify their windmill design based on new understandings and prior data collected. After thorough testing, they can explain how their modification of windmill blade design increased or decreased the amount of power it produced.

Bundling PEs	Connections to Classroom Activity
HS-PS3–1: Create a computational model to calculate the change in the energy of one component in a system when the change in energy of the other component(s) and energy flows in and out of the system are known.	• Student's baseline experiences with windmills to learn about different types of energy and energy transformations is a great building block to deeper explorations of energy. Students can easily use their knowledge from the windmill experiences to think on a quantitative level about energy transformations in a closed system. Students could calculate the relationship between input and output energy.

(NGSS Lead States 2013)

Conclusion

The importance of teachers connecting the model lessons and the NGSS is a major theme of this book. The hard work of translating the model lessons to the NGSS is a powerful professional development strategy that takes teachers' practice to new levels. Many teachers have commented that using the model lessons and reflecting on the connection to the NGSS has helped to teach them quite a bit about learners and learning, the instructional strategies that can improve student achievement and motivation, and has deepened their content knowledge. The result is that teachers have developed their ability to translate the NGSS into practice and know how to approach instructional design from a more research-based perspective.

References

Bybee, R. 2012. Scientific and engineering practices in K–12 classrooms: Understanding a framework for K–12 science education. *Science Teacher*, *78*(9), 34–40.

Duschl, R.A. 2012. The second dimension: Crosscutting concepts. *Science Teacher*, *79*(2), 34–38.

National Research Council. 2012. *A framework for K-12 science education: Practices, crosscutting concepts, and core ideas.* Washington, DC: The National Academies Press.

NGSS Lead States. 2013. *Next Generation Science Standards: For states, by states.* Washington, DC: National Academies Press. www.nextgenscience.org.

3

Model Lesson 1: How Thin Is Foil?

Applying Density to Find the Thickness of Aluminum Foil

James Concannon

About This Lesson

In this activity, I show how high school students apply their knowledge of density to solve an unknown variable, such as thickness. Students leave this activity with a better understanding of density, the knowledge that density is a characteristic property of a given substance, and the ways in which density can be measured.

Is it possible to measure how thick aluminum foil is by using a standard ruler? Of course not! Just imagine the looks you will get from your students when you tell them that they will need to measure the thickness of foil! This is a great activity for ninth-grade physical science students to apply their knowledge of density. This activity could be used after students have a conceptual understanding of density but may not yet know how to apply the concept of density to solve a problem. Hila Science Videos's (2010b) Density and Buoyancy, found at www.youtube.com/watch?v=VDSYXmvjg6M, explains density and buoyancy both quantitatively and conceptually.

In this lesson, students are applying the necessary abilities to do scientific inquiry and to understand the structure and properties of matter (National Research Council 1996). Students are engaged in scientific inquiry in order to work together in developing a procedure and communicating their ideas to solve a problem. By the end of the lesson, students should be able to understand that volume is expressed both as milliliters and centimeters cubed, to know how to use an object's density to solve unknown variables, and to understand that the density of a given material does not change with the shape of the material.

The key to doing this activity is teacher preparation. Each group should have the following materials: aluminum foil, aluminum shot, water, scissors, a graduated cylinder, access to a measuring scale, a ruler, and aluminum wire (an alternative to aluminum shot). Students brainstorm how they could measure the thickness of aluminum foil after the teacher proposes the following driving question: With the materials in front of you, develop a procedure to measure the thickness of a piece of aluminum foil (see Figure 3.1). What is the thickness of the aluminum foil?

Generally, I allow as much time as they need to do this because it is an opportunity for students to practice an essential feature of inquiry: developing a procedure to solve a problem. It is likely that students will have some

Figure 3.1 I show a piece of aluminum foil while I propose the driving question

difficulty coming up with a procedure. Inevitably, some groups will try to only use a ruler to measure the thickness, only to find out that it cannot be done. If the class has made several unsuccessful attempts to solve the problem, I write on the board: (1) density = mass/volume; (2) mass is measured in grams; (3) volume is measured in milliliters or centimers cubed (cm^3); (4) the volume of a solid can be measured by how much water it displaces; and (5) a centimeter cubed is the same as a cubic centimeter (cc).

To solve this problem, students must first calculate the density of aluminum. This is achieved by using water, aluminum shot or wire, and a graduated cylinder. Students take a portion of aluminum shot. Students need to measure the mass of the shot to calculate density. Next, students need to fill their graduated cylinder with a known amount of water. Students then place the entire mass of the shot into the graduated cylinder and determine the water displacement. The density of the aluminum shot is calculated by dividing the mass of the shot (grams) by the amount of water it displaced (milliliters). This calculation should be close to 2.698 g/ml. Aluminum shot can be purchased through a science supplies company. Alternatively, aluminum wire can be purchased at a hardware store.

Next, students need to calculate the density of the aluminum foil. To do this, they measure the mass of the foil (Figure 3.2), the length of the foil

Figure 3.2 Measuring the mass of aluminum foil

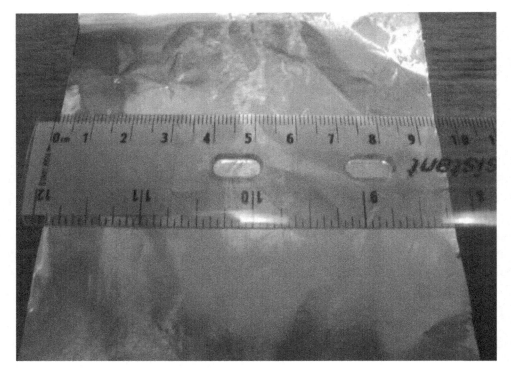

Figure 3.3 Measuring 9.5 cm for the length of the foil

(9.5 cm; see Figure 3.3), and the width of the foil (10.5 cm; see Figure 3.4). Once these measurements are taken, there is only one unknown to determine its density: the thickness of the foil (cm). Since students previously solved the density of the aluminum shot, they can use this calculation of density to solve the unknown thickness of the foil.

Students' calculation of the aluminum foil's thickness may vary slightly from group to group. As a class, we review how to calculate the thickness (see Table 3.1). After determining the thickness of the foil, the class watches a clip from an episode of *Beakman's World* called "Beakman Explains Gas Density." The humorous clip does a great job in elaborating the concept of density to gases.

A short video by Hila Science Videos (2010a), titled Archimedes' Principle, found free (low resolution) at www.youtube.com/watch?v=eQsmq3Hu9HA, can be used to formatively assess students' understanding of density. The beginning of the clip establishes the problem: "A rectangular barge 30 cm long and 200 cm wide floats on the edge of a freshwater lake. A horse jumps into the barge, and the barge sinks 12 cm. How much does the horse weigh?" I stop the video clip at 37 s to give students some time to solve the problem. I ask

Figure 3.4 Measuring 10.5 cm for the width of the foil

Table 3.1 Steps for Calculating the Thickness of the Aluminum Foil

Step	Calculation
1	Density of Aluminum $= 2.698$ g / mL or expressed as 2.698 g / cc.
2	$$\frac{2.698 \text{ g}}{1cc} = \frac{\text{mass of the aluminum foil (g)}}{(\text{length} \times \text{width} \times \text{thickness})}$$
3	$$\frac{2.698 \text{ g}}{1cc} = \frac{0.434 \text{ g}}{(9.5cm \times 10.5cm \times \text{unknown thickness})}$$
4	$1cc \times 0.434$ g $= 2.698$ g $\times (99.75 \text{ cm}^2 \times \text{unknown thickness})$
5	$$\frac{1cc \times 0.434 \text{ g}}{2.698 \text{ g}} = (99.75 \text{ cm}^2 \times \text{unknown thickness})$$
6	$.161cc = 99.75 \text{ cm}^2 \times \text{unknown thickness}$
7	$$\frac{161 cc}{99.75 \text{ cm}^2} = 0.0016 \text{ cm or } 1.6 \times 10^{-3} \text{ cm thick}$$

them not only to calculate the mass of the horse, but also to write an explanation about how they solved the mass of the horse. For this exercise, I do not allow students to work in groups. To ease students' fear of getting a bad grade, I explain that the assessment is not for a grade. After 5 min I ask students to pass their papers to me. Then I click on the play button, and we watch the remaining part of the video. I review students' responses as the remainder of the video plays. Following the video, I open the floor for students to ask questions. Having read some of the students' responses during the video, I can better focus my answers to address their possible disconnections.

Acknowledgements

I would like to thank Mr. Terry Brooks, high school chemistry teacher at Hickman High School in Columbia, Missouri, for showing me this great activity during my student-teaching experience.

References

Beakman explains gas density. 2007. *Beakman's world*. Video. Culver City, CA: ELP Communications, Columbia Pictures Television, Universal Belo Productions, and Columbia TriStar Television Distribution. www.youtube.com/watch?v= w7G4o2alLww (accessed August 5, 2010).

Hila Science Videos. 2010a. *Archimedes' principle*. www.youtube.com/watch?v= eQsmq3Hu9HA (accessed November 28, 2017).

Hila Science Videos. 2010b. *Density and buoyancy*. www.youtube.com/watch? v=VDSYXmvjg6M (accessed November 28, 2017).

National Research Council. 1996. *National science education standards*. Washington, DC: National Academy Press.

4

Model Lesson 2: An Interdisciplinary Theme

Topographic Maps and Plate Tectonics

James Concannon and Linda Aulgur

About This Lesson

This is an interdisciplinary lesson designed for middle school students studying landforms and geological processes. Students create a two-dimensional topographic map from a three-dimensional landform that they create using clay. Students then use other groups' topographic maps to recreate landforms. Following this, students explore some basic ideas about how landforms take shape and how they can change over time. As students work through three distinct learning-cycle phases of concept exploration, introduction, and application, they use art, language arts, and mathematical skills to strengthen or form new science and social studies concepts.

Materials

- Fishing line
- Paper

- Pencils/Pens
- Pie pans
- Colored markers
- Clay
- Ruler/Tape measure
- Pictures of landforms
- Water
- Scissors

National Science Education Standards

Lithospheric plates on the scales of continents and oceans constantly move at rates of centimeters per year in response to movements in the mantle. Major geological events, such as earthquakes, volcanic eruptions, and mountain building, result from these plate motions (National Research Council 1996, 160).

National Council for the Social Studies Standards

Teacher Expectations: Encourage learners to construct, use, and refine maps and mental maps; calculate distance, scale, area, and density; and organize information about people, places, regions, and environments in a spatial context (National Council for the Social Studies 2002, 23).

Introduction

In today's middle schools, it is common to see some level of integration among subjects. Usually, integration begins with a particular theme or content that is relevant in two or more content areas (Trowbridge and Bybee 1996). For example, science teachers can integrate the history of science in their science classes (Trowbridge and Bybee 1996). Due to the nature of this topic, a teacher can easily integrate aspects of social studies into his or her middle school science curriculum.

This lesson was developed through a constructivist lens. A constructivist would agree that the acquisition of knowledge is much more complex than a simple transfer of facts from teacher to student. Students are not like "blank slates" waiting for teachers to write on them. This is because students' prior experiences shape how they have come to understand the world (von Glasersfeld 1995). Accordingly, instruction should not be "pouring information into an empty bucket." The "bucket" that a teacher might be trying to fill may already be full. "Knowledge creation is viewed as occurring through a complex interplay between pre-existing knowledge and new information gathered through interaction with the external world" (Siegler and Ellis 1996, 212). Constructivists believe that learning is an active process and that teachers are guides facilitating the process.

William C. Phillips (1991) has identified several student misconceptions about rocks, mountains, and geological processes. Students entering the classroom may believe that continents do not move and that mountains were created rapidly. Since students come to class with several different conceptions of the world around them, it is important, as a part of instruction, to assess students' misconceptions using a diagnostic assessment. The diagnostic assessment, or pretest, should focus on researched misconceptions. The diagnostic assessment should also include open-ended questions requiring students to write a short response. The diagnostic assessment can be used in the first phase of the learning cycle. The learning cycle was first introduced by Myron Atkin and Robert Karplus (Trowbridge and Bybee 1996). It consists of three phases: concept exploration, concept introduction, and concept application.

Concept Exploration

At the beginning of the lesson, I pass out pictures of different landforms created by the movement of tectonic plates. These pictures are of the (1) Andes Mountains, (2) Himalayas, and (3) Mid-Atlantic Ridge. The students circulate the pictures around the room until everyone has had a chance to look at them. After I collect all the pictures, I hold up one picture at a time, and I ask *driving questions* about them. For example, I ask students to write about how the landforms were created and how long they think it took them to form. The purpose of these questions is to get students thinking about more than just the shape of the landform. After going through the pictures and asking questions, I ask students to turn in their responses to me. This allows me to evaluate their ideas at a later time. When I examine students' responses, I think of ways I might address some of their incorrect ideas in follow-up activities.

Before I pass out the materials, I explain to the class that they will be creating a miniature version of a landform using clay. I tell them that the landform can look something like what they saw in the pictures that were passed around the classroom; however, it is ultimately their decision how they want their landform to look. A student may want his or her landform to look like a mesa (Figure 4.1), butte, or any shape of mountain (Figure 4.2), and that is perfectly fine. After explaining what the students will create, I pass out a portion of clay and a few pie pans with a small amount of water for students to

Figure 4.1 A student's representation of a mesa

dip in their hands to make the clay easier to mold. The pie pan works well for students to get their entire palms wet. Once students have had a chance to create their landform, I ask this driving question: "The landform you have built is a three-dimensional model, the dimensions being height, width, and depth. What is a way you could represent the three dimensions of your model landform on a two-dimensional piece of paper?" I then add, "You can do this with the materials I am going to pass out to you." I then pass out rulers, fishing line, paper, and pencils.

I give students some time to think about the problem and test their ideas with the materials. I then ask students to summarize their procedure, and if they think their procedure would work well if I were a climber and trying to figure out the best way to climb up the landform from looking at the maps they created. Generally, students think of drawing the profile of the landform, like in Figure 4.2, from three or four different sides. Some students will think that drawing their landform from four different sides is a great idea. I tell students to pass their pictures to the person next to them. After students pass

Figure 4.2 A student's representation of a mountain

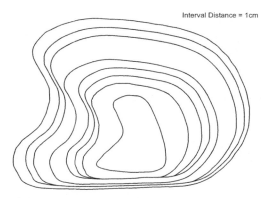

Interval Distance = 1cm

Figure 4.3 Generic topographic map

their pictures, I ask, "Which side of the landform is steeper, taller, or wider? Tell me which side is smoother, and which side has more bumps." From just looking at the pictures, the students will not know. I then give the students a piece of the puzzle to help them in answering the driving question. I pass out a generic topographic map, similar to Figure 4.3.

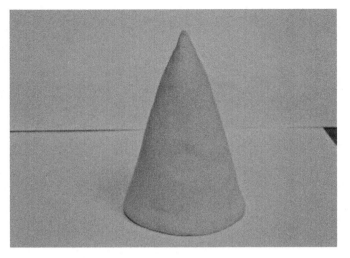

Figure 4.4 Model landform

After I pass out the map, I ask students to discuss with their neighbor what they think the map is trying to represent. Within three minutes, I ask, "So what do you think this map is showing us?" Generally, I explain how to interpret the map using a series of questions if students have not figured it out. The main ideas of interval and contour lines are introduced. After students have an idea as to how to interpret the map, I ask them: (1) "Which side of this landform is steepest?" (2) "How do you know?" (3) "What does it mean if there is little or no gap between the contour lines?" and (4) "Which side is the steepest?" I then tell the class that they will make a topographic map similar to this one using their model landform (Figure 4.4).

Concept Introduction

I pass out the fishing line and rulers. I also pass out a piece of copy paper to each student. I find that the copy paper works better since it has no lines. I tell the students to put their model landform (Figure 4.4) on their piece of paper. I then give students the following directions:

1. Stand the ruler on its end, with the 0 cm mark at the base of the model. The ruler should be perpendicular to the desk; it should not be at an angle along the model (Figure 4.5).
2. At every 1 cm increment along your landform, make a notch in the clay with a pencil (Figure 4.6).

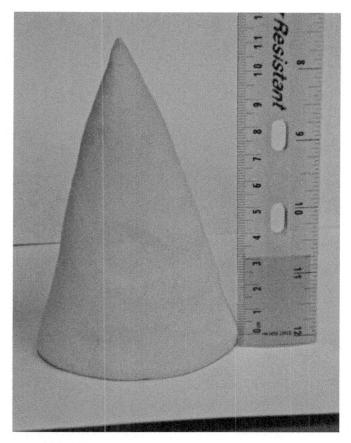

Figure 4.5 Measuring the landform

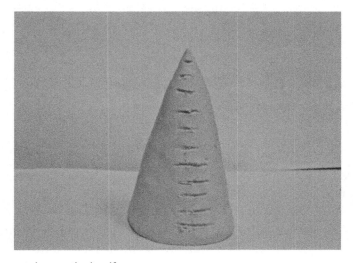

Figure 4.6 1 cm notches on the landform

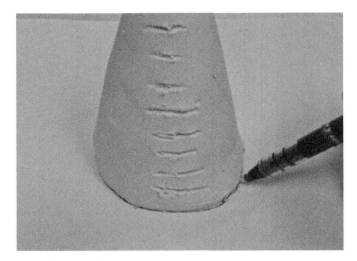

Figure 4.7 Drawing the first contour line at 0 cm

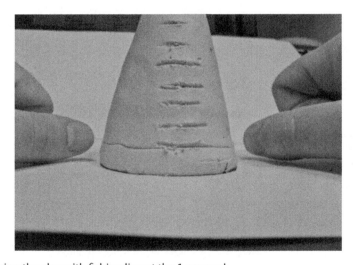

Figure 4.8 Slicing the clay with fishing line at the 1 cm mark

3. With a pen, trace along the perimeter of your model (Figure 4.7). This is the 0 cm interval.
4. Using fishing line, cut a cross-section of your model at the 1 cm mark (Figure 4.8). Do not remove the bottom half of the model from the paper. Place the top portion of the model back on the bottom portion in such a way that they "fit" back together.

Figure 4.9 Aligning the landform in one direction by making a mark on the paper

5. Under the notched intervals, mark the paper by drawing a line. This line is to correctly align the model after each slice (Figure 4.9).
6. Remove the bottom portion of the model (Figure 4.10).
7. Center the top portion of the model inside the 0 cm interval. Correctly align the notches on the model with the mark on the paper. Trace around the base of the model. This is the 1 cm interval (Figure 4.11).
8. Repeat steps 4 to 8 for each notch on the model (Figure 4.12).

After they create their topographic maps, I ask the students to clump the clay back together. I then tell them to pass their topographic maps to the person next to them. With the topographic map, ruler, and clay, the neighboring student re-creates the model landform. After students do this activity, I ask how they knew that their neighbor's model landform looked something like what they had molded. The primary purpose of this activity is for the students to be able to explain to me why their neighbor's landform looked similar to the one they re-created. In this conversation, I reinforce the purposes of contour lines and interval distances.

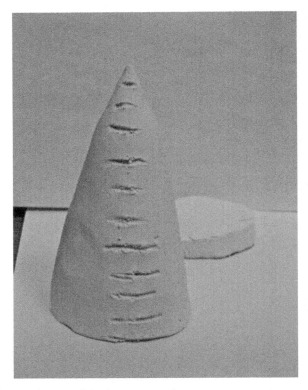

Figure 4.10 Landform with 1 cm of the base removed

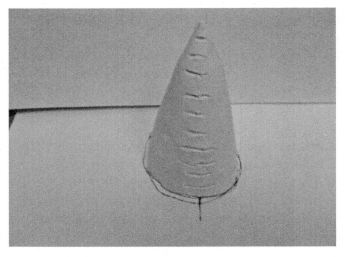

Figure 4.11 Drawing the second contour line at the 1 cm mark

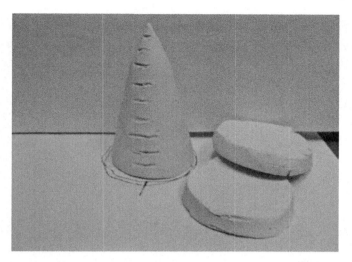

Figure 4.12 The directional mark on the paper, two contour lines drawn, with 2 cm of the base removed

Concept Application

Topographic mapping gives teachers the opportunity to do follow-up activities about plate tectonics and mountain-forming processes that directly relate to the initial driving questions. In this activity, I provide a map (taken from the U.S. Department of Transportation website) of the continents showing the major tectonic plates and their direction of movement (Figure 4.13). I also provide a colored piece of construction paper on which the map barely fits.

I explain to the students that the arrows on the map represent the direction in which the plates are moving. Some plates move away from each other (diverging plate boundaries), and some are moving into each other (converging plate boundaries). I tell students to cut along the plate boundaries with an X-ACTO knife or scissors. After cutting the plates apart, I tell students to piece the map back together, showing gaps where plates are diverging and overlapping the pieces where the plates are converging.

Although it is shown two-dimensionally, the map is actually a three-dimensional representation. The left portion of the map is continued on the right side and vice-versa. The two-dimensional map is actually showing a sphere, like a basketball, having a fixed volume. The colored paper represents a fixed volume. I tell my students that the Earth does not increase or decrease in volume; it has a fixed volume. Therefore, students must keep all the pieces of the map on the colored paper (Figure 4.14).

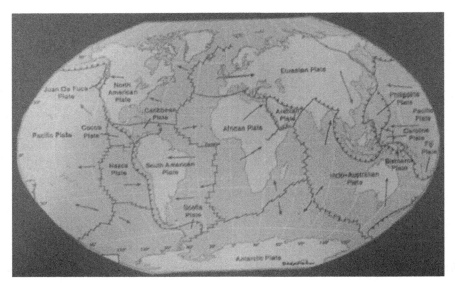

Figure 4.13 Map of Earth's tectonic plates

Map is available in the *Technical Manual for Design and Construction of Road Tunnels—Civil Elements*, Section 13.2.1: Earthquake Fundamental (U.S. Department of Transportation Federal Highway Administration, Figure 13–1 Major Tectonic Plates and Their Approximate Direction of Movement)

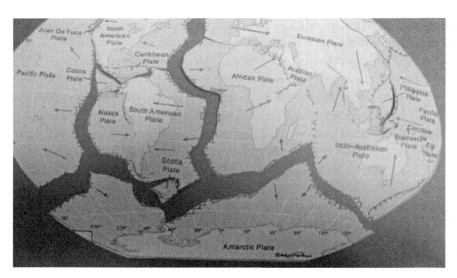

Figure 4.14 Students arranging plates to show where divergent and convergent boundaries are located

Map is available in the *Technical Manual for Design and Construction of Road Tunnels—Civil Elements*, Section 13.2.1: Earthquake Fundamental (U.S. Department of Transportation Federal Highway Administration, Figure 13–1 Major Tectonic Plates and Their Approximate Direction of Movement)

Next, I have students write down what they think is causing the plates to move and what types of landforms are being created at the different plate boundaries. Rather than explaining the geological process, I bring my students to the computer lab. In the computer lab, students can either do a Google or Yahoo video search for "The Early Earth and Plate Tectonics," which is a short, five-minute clip provided by National Geographic. The clip is free, easily accessible, and engages students using colorful animation to show plate-tectonic processes. Since the students are already in the computer lab, I ask them to research the follow-up questions: (1) What mountain range is located at the Nazca and South American plate boundary, and how did this mountain range form? (2) What mountain range is located at the Indo-Australian and Eurasian plate boundary, and why did this mountain range form? (3) What mountain range is located between the Juan de Fuca and the North American plate boundary, and why did this mountain range form? (4) What is the Great Rift Valley, where is it located, and why was it formed? and (5) What is the Mid-Atlantic Ocean Ridge, and why was it formed? After students research these questions and bring their answers back to class, we have a discussion about different plate boundaries and the landforms that are created via tectonic processes. Before finishing the lesson, I write the driving questions back on the board, and I have students write their answers on a separate sheet of paper. Doing this, I assess students' post-instruction ideas, reading for continued misconceptions within their responses.

Conclusion

At the end of this lesson, students have successfully constructed new concepts in both science and social studies. During the learning process, students have had the opportunity to use writing, as well as art and math skills, to arrive at content knowledge in the areas of mapping and topography. Activities such as the one described give the learner a chance to actively construct important concepts tied to national standards. Using the three phases consisting of concept exploration, introduction, and application, the students were able to work through their misconceptions and come away with a much deeper level of geographical knowledge.

Teachers can set the stage for active constructivist learning in designing activities in which students are being asked to predict and refine the investigative processes. Activities that include clear concept investigation lead to students developing their higher-order critical thinking skills in a constructivist setting.

References

National Council for the Social Studies. 2002. *National standards for the social studies.* www.socialstudies.org/ (accessed November 28, 2017).

National Geographic. 2010. The early earth and plate tectonics. http://vodpod.com/ watch/2799255-the-early-earth-and-plate-tectonics (accessed May 28, 2010).

National Research Council. 1996. *National science education standards.* Washington, DC: National Academies Press.

Phillips, W.C. 1991. Earth science misconceptions. *The Science Teacher, 58*(2), 21–23.

Siegler, R., and S. Ellis. 1996. Piaget on childhood. *Psychological Science, 7,* 211–215.

Trowbridge, L., and R. Bybee. 1996. *Teaching secondary school science: Strategies for developing scientific literacy.* Englewood Cliffs, NJ: Prentice-Hall.

U.S. Department of Transportation Federal Highway Administration. 2009. Figure 13–1: Major tectonic plates and their approximate direction of movement. In *Technical manual for design and construction of road tunnels—Civil elements.* www.fhwa.dot.gov/bridge/tunnel/pubs/nhi09010/13.cfm (accessed May 28, 2010).

von Glasersfeld, E. 1995. *A way of knowing and learning.* London: Falmer Press.

5

Model Lesson 3: Students' Use of the PSOE Model to Understand Weather and Climate

Patrick Brown and James Concannon

About This Lesson

One tried-and-true way to hook students' attention and promote long-lasting understanding is to sequence science instruction in an explore-before-explain instructional sequence. In these lessons, students investigate the interaction between "cold" and "hot" substances and density to understand how fluids interact in air (some weather phenomenon) and water (ocean currents). The student-based activities are suited to a research-based strategy called the PSOE (predict, share, observe, and explain) sequence of instruction.

A thoughtfully sequenced Earth and space science demonstration can be fundamental to helping students formulate, revise, and develop ideas to promote long-lasting understanding. We have success teaching sixth grade students using an exploration-before-explanation instructional sequence called the PSOE model that consists of the following phases: predict, share, observe, and explain (Haysom and Bowen 2010; Stepans 1996). The PSOE instructional sequence is a useful tool for designing science lessons because it helps teachers to focus on important concepts and highlights that students learn best when they are actively engaged in thinking and doing and have the chance to build new ideas before teacher explanations (Bybee 1997). The *Predict* stage captivates students in the lesson and allows teachers to identify students' initial conceptions (including misconceptions). The *Share* stage is a time for students to collaborate, reformulate, and refine scientific ideas. The *Observe* stage presents students with firsthand experiences with quantitative or qualitative data and evidence. Finally, the *Explain* stage allows students to generate scientifically accurate ideas based on data they have collected or observed during the demonstration. Using this approach, teachers can help students identify, redefine, elaborate, and change their initial conceptions through self-reflection and interaction with their peers and observations.

We derived the investigation using Stepan's (1996) description of the phases and purposes of the POE model to include the three-dimensional nature of the K-12 Frameworks (seamless integration of essential science practices, crosscutting concepts, and disciplinary core content). The use of the POE helped us prompt sixth grade students enrolled in Earth and Space science class to think about the mechanism behind weather patterns. Weather and climate are highlighted in the *Next Generation Science Standards.* The second through sixth grade band emphasizes "Variations in density due to variations in temperature and salinity drive a global pattern of interconnected ocean currents" (MS-ESS2–6, NGSS Lead States 2013) (see Table 5.1 for Standards connections). Understanding the relationship between "cold" and "hot" substances and density is important for understanding how fluids interact in air (some weather phenomena) and water (ocean currents). We use what we term "molecular talk" during discussions to bridge students' observations of science on the macroscopic level with a microscopic explanation of weather and ocean currents.

Table 5.1 Connections to Standards

Standard
MS-ESS2–6. Earth's Systems

Performance Expectation
MS-ESS2–6. Develop and use a model to describe how unequal heating and rotation of the Earth cause patterns of atmospheric and oceanic circulation that determine regional climates

Science and Engineering Practices
Developing and Using Models

Disciplinary Core Ideas
ESS2.C: The Roles of Water in Earth's Surface Processes

Crosscutting Concept
Systems and System Models

The lesson that follows is a quick and cost-effective way to prompt sixth grade student thinking about the mechanism behind weather patterns.

Materials and Safety

- Cold water (approximately (9°C)
- Two 50 ml Erlenmeyer flasks
- Two 900 ml beakers
- Red food coloring
- Blue food coloring
- Hot water (approximately 65°C): Students should handle hot water with caution and use protective aprons, gloves, and goggles

Demonstrations 1 and 2: Hot Water Set into Cold Water, Cold Water Set into Hot Water (Seven Minutes)

We have students predict (predict phase) what happens in two different set-ups (see Figures 5.3 and 5.4 below). In the first set-up, hot water (65°C) is placed in a 50 ml Erlenmeyer flask and colored red. Then, students are told that the flask will be submerged into a 900 ml beaker filled with cold (9°C) water. The second set-up is the exact opposite: students are told that the 50 ml flask with cold water (9°C) dyed blue will be placed in a 900 ml beaker with hot water (65°C). (Note: The food coloring allows students to identify the differences in the temperatures of water. We decided to only dye the water in the flask in each scenario so that it was easier to see when changes occurred.) After telling students the scenarios, we have them draw on paper their predictions for both scenarios to show what will happen to the water in the flasks after being submerged into the beaker (see Figures 5.1 and 5.2 for students' predictions).

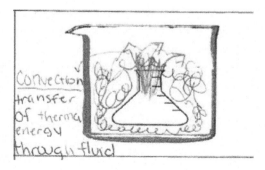

Figure 5.1 Student prediction for the transfer of thermal energy from "hot" to "cold"

Figure 5.2 Student prediction for the transfer of thermal energy

Figure 5.3 Picture of the transfer of the thermal energy from "hot" to "cold"

Next, during the share phase, students talk through their ideas with a partner. Students' sharing included all imaginable possibilities. Some students thought that in both set-ups the dyed water (red or blue) would move out of the flask into the beaker of water. Other students believed that only the red-colored or blue-colored water would move into the beaker of water. A few students thought that nothing would happen and neither of the colors of water (red or blue) would move. During this time, students asked each other questions, requested evidence for claims, and explained their thinking using logical reasoning. To our surprise, some students discussed that they thought water could only move in one direction in each set-up: (from the

Figure 5.4 Picture of "cold" water dyed blue not moving into less dense, "hot" water

flask to the beaker), reasoning that water could only move into the beaker because the flask was already full. The concept that water could only move one way became a very appealing idea for the class, which led to performing an additional demonstration (see Demonstration 3). (See Figures 5.1 and 5.2 for student predictions.)

Teachers should remain "active" listeners during the discussion, encourage student-to-student conversations, and not focus on correct answers. Our encouragement during the share phase was aimed at motivating students to talk through their ideas and not at whether they had accurate conceptions.

Once all students had shared their ideas, it was time to have them *observe* what would happen when two different temperatures of water were allowed to come in contact with each other. The food coloring was pivotal for promoting conceptual understanding and provided the visual support for students to be able to explain the phenomenon. First, we put the 50 ml flask containing hot water (65°C) dyed red in the 900 ml beaker containing cold water (9°C). The red-colored water quickly moved in a steady stream into the surrounding cold water (9°C). Students said that the first set-up looked like a "volcano exploding" (see Figure 5.3).

Next, we placed the cold (9°C) dyed water in the 50 ml flask into the 900 ml beaker containing hot water (65°C). To students' surprise, the cold water did

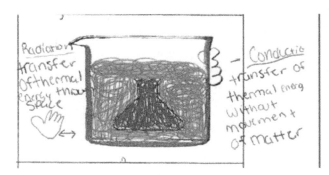

Figure 5.5 Student drawing and explanation of "cold" water dyed blue not moving into less dense, "hot" water

not move into the warm water in a steady stream (see Figure 5.4). In fact, more than a minute after the flask had been placed in the beaker, the cold water still had not moved into the beaker.

At the end of the *Observe* phase, students drew their observations. Students observed that the water moved from hot to cold in the first set-up. In addition, students witnessed that the cold water did not move into the surrounding hot water (see Figure 5.5). Students' observations were not graded, but helped them to construct evidence-based claims that occurred later in the lesson.

Demonstration 3: Hot Water Set into Cold Water

We had not anticipated the idea that water could only move in one direction—from the flask to the beaker—and students determined that we needed to test another set-up. In the third set-up, we filled a flask with hot water (65°C) and placed it in a beaker filled with cold water dyed blue. When we placed the hot water (65°C) in the flask into the cold blue water (9° C) in the beaker, some students were surprised by the results. After a minute, we removed the flask from the beaker and students observed cold water slowly moving in a steady stream to the bottom of the flask. As one student explained, the steady stream looked like a "reverse volcano."

Discussion: Connecting Key Concepts and Introducing New Terms

Teachers can tap into children's innate curiosity to help them develop more complex thinking skills in which firsthand science experiences serve as the foundation for deeper learning. The skill of asking children to make connections between firsthand experiences and science content is an important step in developing conceptual understanding and can be facilitated through purposeful and focused investigations. The missing information needed to help develop students' understanding of some weather phenomena was related to the density of "hot" versus "cold" water. Thus, once students explained their observation, we challenged them to make a scientific claim about the density of hot vs. cold water, the interactions that take place when hot and cold materials come into contact with each other, and to propose a mechanism for which some large air masses and water masses move on Earth. The explanation following the exploration was a way for students to develop a deeper conceptual understanding by anchoring new ideas with their firsthand experiences and observations. In addition, the evidence-based claims are a way for students to use model-based reasoning (i.e., crosscutting concept) to explain a phenomenon (weather and ocean currents) that is not easily observable firsthand.

Prior to the demonstration, students had learned about density when studying properties of matter. To guide the discussion, we asked, "Do hot water and cold water have different chemical compositions?" This is important to ask because it helps point out that the only factor affecting density is temperature, not a change in chemical composition. Next, we had students write claim statements about the density of hot and cold water based on observations. For example, students wrote for their evidence-based claim that, "the hot water was less dense and floated to the top," and "the cold water was more dense and sank to the bottom." Students supported their claims with evidence by describing what they observed during the demonstrations. One student wrote that during the demonstration, "The hot water looked like a cloud on top of the cold water." Another student explained that the cold water "dropped to the bottom" and looked like a "steady flowing blue stream of water."

We learned that even though we had students make evidence-based claims, not all students have a scientific understanding of density. As a result, our students benefited from what we term "molecular talk," where we compare high and low densities for a material of the same volume. We used ping pong balls to represent water molecules and showed that in high densities the particles are packed tightly together with little space between them, whereas in low densities particles are more loosely arranged with space between them. Thus, students were able to explain the evidence (e.g., "hot" water dyed red floats and "cold" water dyed blue sinks) that supports the claim (cold water is denser than hot water). In addition, the "molecular talk" conversation helped prepare students to use scientific reasoning to connect their claims and evidence (termed a "warrant" or "reasoning" statement). Students described that cold water sinks because the "molecules in cold water are more tightly packed together than water molecules in 'hot' water." We liked that our students used warrant statements to support their claims based on evidence, an important skill advocated by the Common Core State Standards (CCSS) for argumentative writing in English language arts (National Governors Association Center for Best Practices [NGA] and Council of Chief State School Officers [CCSSO] 2010). Thus, the use of a claims, evidence, and warrant statement structure for writing is a way to make learning more interdisciplinary.

The most difficult part of students' evidence-based claims is proposing a mechanism for which large air masses or water masses interact over time. Students can relate what they observed firsthand with the hot and cold water demonstrations and density to weather. Students reason that the Earth's atmosphere and oceans are made up of different air masses or water masses with varying temperatures (prior knowledge) that interact. For example, when air masses or water masses with different densities collide, the denser air mass or water mass sinks below the less dense air or water mass. In this way, the different color water demonstrations provide visual support for the interaction of air masses and represent the interaction of air or water masses as a result of the interaction of different temperatures.

The *Explain* phase was also a chance for students to derive concepts and new terminology from firsthand experiences. For example, we called the movement where cold surface water sinks below the warmer underlying water "convection." By introducing new formal terminology and concepts in light of students' firsthand experiences with a visual model, we maintained high levels of engagement during discussions and students were able to attach meanings to these new ideas.

Conclusion

We found that these activities increased our sixth grade students' interest and helped them learn about convection currents in water and air. Although convection demonstrations have been performed by many teachers for a long time (see Roth n.d.), the use of the PSOE demonstration was an exciting way to promote science understanding and reasoning to propose explanations for challenging content (Haysom and Bowen 2010; Stepans 1996). Using different colored waters has many applications for teaching process and interactions. The concept of convection driven by temperature and density differences can be used to explain many Earth and space science concepts such as: (1) Mantle convection playing an important role in plate tectonics;

(2) formation of cumulus clouds and thunderstorms; (3) separation between layers in the Earth's (and other planetary) atmospheres; and (4) convection and granule development at the solar surface. In addition, having students describe their understanding of the demonstration and discussion through writing helps prepare them for advanced science topics and develops their thinking and acting like scientists (supporting and proposing ideas through writing). Most importantly, the exploration-before-explanation sequence promotes deep conceptual understanding because new ideas and concepts are based on students' firsthand experiences with data and evidence.

References

Bybee, R.W. 1997. *Achieving scientific literacy: From purposes to practices.* Portsmouth, NH: Heinemann Educational Books, Inc.

Haysom, J., and M. Bowen. 2010. *Predict, observe, explain: Activities enhancing student understanding.* Arlington, VA: NTSA Press.

National Governors Association Center for Best Practices [NGA] and Council of Chief State School Officers [CCSSO]. 2010. *Common Core State Standards (English Language Arts).* Washington, DC.

NGSS Lead States. 2013. *Next Generation Science Standards: For states, by states.* Washington, DC: National Academies Press. www.nextgenscience.org/next-generation-science-standards (accessed September 28, 2015).

Roth, J. n.d. Lesson plan: Hot, cold, fresh, and salty. NOAA Ocean Service Education, Lesson Plan Library. http://oceanservice.noaa.gov/education/lessons/hot_cold_lesson.html (accessed June 28, 2015).

Stepans, J. 1996. *Targeting student's science misconceptions: Physical science concepts using the conceptual change approach.* Riverview, FL: Idea Factory Inc.

6

Model Lesson 4: Teaching Bernoulli's Principle through Demos

Patrick Brown and Patricia Friedrichsen

About This Lesson

One proven strategy to help students make sense of abstract concepts is to sequence instruction so that students have exploratory opportunities to investigate science before being introduced to new science explanations (Abraham and Renner, 1986; Renner, Abraham and Birnie, 1988). To help physical science teachers make sense of how to effectively sequence lessons, this chapter summarizes our experiences using an exploration-explanation sequence of instruction to teach Bernoulli's principle to prospective middle and secondary science teachers in a science Methods course. We use demonstrations during our Bernoulli's unit to help students go back and forth between their observations of phenomenon and what occurs at the microscopic level, using what we have termed "molecular talk." Students engage in guiding questions, consider their old and new understandings of science, and use evidence to construct new ideas during all stages of the lesson.

In this unit we use easy-to-do, classic demonstrations of Bernoulli's principle to help students go back and forth between their observations of phenomenon and what occurs at the microscopic level, using what we have termed "molecular talk." Students engage in guiding questions, consider their old and new

NSES for Teaching Bernoulli's Principle

- **Content:**

- Content Standard: 9–12, Physical Science. In solids the structure is nearly rigid; in liquids molecules or atoms move around each other but do not move apart; and in gases molecules or atoms move almost independently of each other and are mostly far apart.
- Content Standard: 9–12, Physical Science. Objects change their motion only when a net force is applied. Laws of motion are used to calculate precisely the effects of forces on the motion of objects.

Science as Inquiry:

Understandings about Scientific Inquiry:

- Scientists develop explanations using observations (evidence) and what they already know about the world (scientific knowledge). Good explanations are based on evidence from investigation

Teaching Standards:

- Teachers of science plan an inquiry-based science program for their students.

  - Select teaching and assessment strategies that support the development of student understanding and nurture a community of science learners.

- Teachers of science guide and facilitate learning. In doing this, teachers orchestrate discourse among students about scientific ideas.

understandings of science, and use evidence to construct new ideas during all phases of the lesson. Many of the activities we use are adapted from Stepans (1992) and meet a number of the *National Science Education Standards* [NSES] (1996) (see Sidebar 1) by emphasizing improving science literacy through active, student-centered learning.

The Exploration Activities and Explanations

Activity 1: Straw and Paper Tent Demo

We use a *Predict, Share, Observe,* and *Explain* (PSOE) sequence of science instruction to identify students' ideas and facilitate what Stepans (1996) would call "conceptual change." First, students *Predict* what will happen when I use a straw to blow into a paper tent (see Figure 6.1).

At this point it is critical to have students commit to an outcome and what they think will happen. Students write their prediction on a sheet of paper that is not graded. Next, students *Share* their predictions and provide an explanation with a partner. Many times when students share, they begin to revise and elaborate on their initial predictions. What the teacher will find is that in most cases students predict that blowing through the straw will cause the paper to either move upwards or move backwards (i.e., they hold typical misconceptions about air flow and air pressure and that blowing air on an object will cause it to move away) Then, students test their predictions and *Observe* what happens to the paper tent when I blow into the tent (see Figure 6.2).

Most students are shocked when their predictions are found to be inaccurate. Having

Figure 6.1 Using a straw to blow into a paper tent

Figure 6.2 What happens to a paper tent when it is blown into?

found many of their predictions to be incorrect, students begin to feel uncomfortable with their prior ideas. In groups of two to three students, explain possible explanations for the phenomenon in groups. The outcome of the discussions that occur in small and large groups is that students ask each other questions and articulate their ideas through their explanations. Having students explain the concepts to each other and in large groups helps them to develop a better understanding than what any one student could have developed alone.

At this point in the lesson we were unsure whether everyone understood the phenomenon for the following reasons: (1) There were different predictions (i.e., the tent would collapse and the tent would move up); and (2) the level of their explanations varied (i.e., some students discussed the demo at the macroscopic level and others started talking about how molecules in air interact when a force is exerted upon them). However, because it seemed like many students were ready to talk about the Bernoulli's principle at a molecular level (micro), we used another demo following the PSOE sequence (Stepans 1994). This time we explicitly challenged students to predict, share, observe, and explain what would happen with another Bernoulli's demo at the macroscopic and molecular level using "molecular talk."

Activity 2: Soda Cans Demo
- *Predict*: what will happen when I use a straw to blow between two soda cans resting on straws (see Figure 6.3)?
- *Share*: your predictions and provide an explanation with a partner.
- *Observe*: what happens to the two soda cans when I blow in between them (see Figure 6.4)?
- *Explain*: based on your observations, discuss with a partner the underlining principles of what is occurring to the molecules in air using "molecular talk." Also, draw your explanation using a white board.

At the end of both demos students had "invented" two explanations for Bernoulli's principle using "molecular talk." These two explanations included:

1. Most students thought that blowing between two objects causes the air between the two objects to move in the same direction. Molecules moving in the same direction have less random collisions. As the speed of a moving fluid increases, the pressure within the fluid decreases due to there being less random collisions among air molecules.
2. A few students explained that blowing between two objects (inside the tent's walls or between the soda cans) removes all of the air in that area. Because there are no molecules in the air, the pressure is lower (i.e., inside the paper tent and between the soda cans) than the surrounding

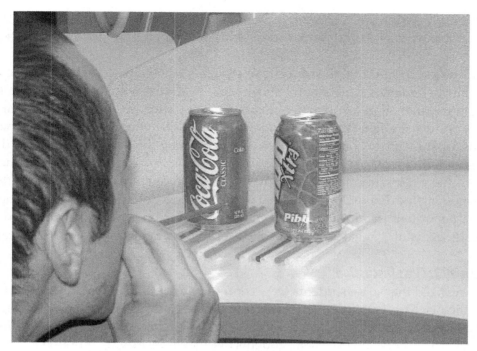

Figure 6.3 Using a straw to blow between two soda cans resting on straws

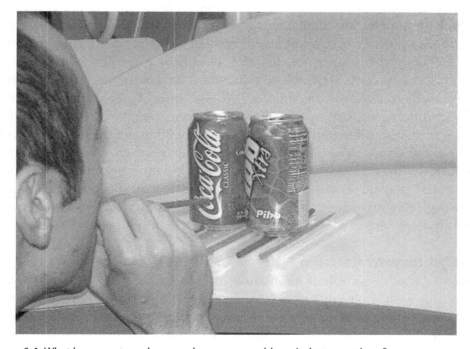

Figure 6.4 What happens to soda cans when someone blows in-between them?

air where air molecules are present and colliding. Surrounding air pressure is high and moves from areas of high pressure to low pressure.

We were surprised that the majority of students who held accurate conceptions of the phenomena (explanation 1 above) did not discuss with their peers the conflicting explanation (explanation 2 above). Thus, after we discussed the demo we conducted a "checkpoint for understanding." The purpose of the checkpoint was for students to commit to an idea by explaining their understanding of the demo. Individually, students wrote down on an index card what they learned from the exploration activities and what they were still confused about after witnessing the demos. We found that many students described the phenomenon (i.e., paper tent flattens and soda cans move towards each other), but had difficulty explaining what is happening to the modules in moving vs. stationary air.

Activity 3: Class Discussion

We realized that some students were confused about Bernoulli's principle based on their two explanations. We focused our whole-class discussion on providing a logical explanation for the two demos. We provided the initial driving question—what is air composed of? Then, students discussed the driving question in small groups before opening the conversation to the whole class.

From prior experiences students knew that air is made up of carbon dioxide, oxygen, nitrogen and other trace gases. Students used their science knowledge to think through the idea that blowing between two objects removes all of the air in that area. For example, students recognized errors in their thinking because air expelled from the lungs contains air molecules (e.g. oxygen, carbon dioxide, etc. . . .). Thus, students were able to refute the idea that the paper tent collapsed and the soda cans moved together due to the absence of air molecules relative to the surrounding air. To move the discussion along, we provided scaffolding by asking students probing questions such as "what is happening to the molecules in the air when we blow through the straw" and "what are molecules doing in the surrounding air?" All students decided that blowing creates a "stream of air molecules" that are all moving in the same direction. At this point, all students were beginning to connect that there are differences between what is occurring with air molecules in a "stream of air" and the "surrounding air."

Activity 4: Computer Simulation

Next, we used a computer simulation to help students understand what is occurring with molecules in a "stream of air." The simulation shows how molecules in the air interact when a force is exerted upon them (available at www.youtube.com/watch?v=8vqMotb6m3c). From the demo, students were able to explain that in a "stream of air" molecules are all moving in the same direction and the collisions among molecules are fewer compared to air molecules in stationary air. In other words, as a flowing fluid gains speed, the internal pressure in the fluid decreases in relation to the surrounding air. (For a mathematical explanation of Bernoulli's principle, see Hewitt, 2004.)

Activity 5: Paper Sheet Student-Demo

After learning about Bernoulli's principle, students were eager to test these new ideas out on their own. The demo described below allows students to feel the effects of air pressure firsthand (Stepans, 1996).

- Get out sheets of paper. Cut off one strip of paper.
- Put one end (5 cm) of the paper between two pages in a book.
- On your own: Predict what will happen when we blow across the paper.
- Check your prediction.
- Explain the discrepant event using molecular talk.

The end result of activities 4, 5, and 6 was that students made the connection between the microscopic and macroscopic phenomena. In summary, students used "molecular talk" to describe that the pressure difference of a "stream of air" (referring to the fluid) and surrounding air is enough to cause the paper tent to collapse, the soda cans to move together, and the strip of paper suspended between two pages of a book to move up.

Activity 7: Ping Pong Ball Demo

The final activity in this lesson has students use these science ideas to solve new problems. We provided students with two small plastic cups and a ping pong ball, and used the PSOE sequence so that students can further test their understanding of Bernoulli's principle in a new situation.

- *Predict:* how you can move the ping pong ball from one glass to the other without touching the ping pong ball or glass? (draw out)
- *Share:* your predictions and provide an explanation with a partner.
- *Observation:* test your prediction.
- *Explain:* write an explanation for your observations using "molecular talk." (Evaluation)

Conclusion

The results of the exploration-explanation instructional sequence help students to understand Bernoulli's principle. Providing experiences before terminology offers a way for teachers to promote critical thinking because students challenge their existing conceptions and carry out meaningful investigations. Additionally, beginning units with demos and discrepant events has the added benefit of engaging student thinking about observable phenomena (Stepans, 1994). Once students are engaged, investigative questions help them to explore these phenomena and make scientific claims based on their experiences. Using demonstrations and computer simulations provides students with visual support that promotes deep learning because new concepts are anchored on preexisting knowledge and experiences. This strategy is consistent with the way people learn, and many students need to make connections between firsthand experiences and observations and new terms, concepts, and scientific ideas (Bransford, Brown, and Cocking 2000). Due to the success of this unit, we would like to provide students with even more real-world elaborations and in the future have them investigate how Bernoulli's principle applies to other real-life circumstances such as the aerodynamic lift that occurs with an air plane wing during flight (see Hewitt, 2004). While sequencing science lessons in an exploration-explanation sequence presents some challenges, it helps students to make connections between their experiences and to make sense of abstract scientific terms and concepts.

References

Abraham, M.R., and J.W. Renner. 1986. The sequence of learning cycle activities in high school chemistry. *Journal of Research in Science Teaching*, 23, 121–143.

Bransford, J., A. Brown, and R. Cocking. 2000. *How people learn: Brain, mind, experience, and school.* Washington, DC: National Academy Press.

Bybee, R. 2002. Scientific inquiry, student learning, and the science curriculum. In R. Bybee, ed. *Learning science and the science of learning* (pp. 25–35). Arlington, VA: NSTA Press.

Hewitt, P. 2004. Bernoulli's principle: Understanding Bernoulli's principle as it applies to aerodynamic lift. *The Science* Teacher, 71, 51–55.

National Research Council. 1996. *National science education standards.* Washington, DC: National Academy Press.

Renner, J.W., M.R. Abraham, and H.H. Birnie. 1988. The necessity of each phase of the learning cycle in teaching high school physics. *Journal of Research in Science Teaching*, 25, 39–58.

Stepans, J. 1996. *Targeting student's science misconceptions: Physical science concepts using the conceptual change approach.* Riverview, FL: Idea Factor Inc.

7

Model Lesson 5: Gravity Is Easy to Understand, Right?

The Difference between Calculating and Comprehending

James Concannon

About This Lesson

Physics instruction looks different from one classroom to the next; however, the outcome of those classrooms should be one and the same. Students should understand simple concepts and be able to transfer concepts to more complex problems. In this lesson, I would like to show you how I provide students with an opportunity to explore their ideas before formulae or calculations are introduced.

As a beginning high school physics teacher, my first thought about teaching gravity was to approach the topic as I was taught. This consisted of formulae, example problems, knowing which formula to use for a given problem, and finding a way to get the correct answer. Physics was simply a body of knowledge consisting of formulae to help me solve any kind of calculated problem. It was not until my mentor encouraged me to read a book titled *Conceptual Physics*, by Paul G. Hewitt, that my ideas about how to teach physics began to shift. Since then, I have come to find out that although the formulae are nice, teaching students how to use formulae does not constitute teaching physics. So, what is "teaching physics"?

Physics instruction looks different from one classroom to the next; however, the outcome of those classrooms should be one and the same. Students should have a deep understanding of simple concepts and be able to transfer those concepts to the bigger picture (National Research Council 2005). All too often, teachers lecture about how to use formulae, resulting in students learning how to use physics without fully understanding the fundamental concepts. In such cases, students' knowledge of physics concepts is relative to how familiar they are with the problem they are trying to solve. Rather than creating the situated learning, physics teachers should facilitate students' fluid understanding of physics concepts.

How is this done? Put down the calculator, at least initially, and hold off opening the traditional physics textbook. Think of a simple demonstration, maybe a ball rolling through an inclined tube. Ask students to predict how changing one variable affects another variable. For example, how does changing the angle of the tube affect the time it takes for the ball to get to the bottom? Ask students for their predictions before showing them the demonstration. Some students' predictions could be correct. After the demonstration, ask students for their observations and explanations as to why one variable may or may not have had an effect on the other variable. The explanation is essential. In the explanation, students are asked to reveal their ideas, which are often incorrect. Unfortunately, sometimes teachers eliminate opportunities for students to think by prematurely providing their class with formal explanations of what happened and why. Students need opportunities to think about the situation and consider what caused specific outcomes. Do not give students the answers; make them think. In essence, initially take a conceptual approach to teaching physics concepts.

National Science Education Standards

Grades 9–12 Physical Science Content Standard B: Motion and Forces: The magnitude of the change in motion can be calculated using the relationship $F = ma$, which is independent of the nature of the force. Whenever one object exerts force on another, a force equal in magnitude and opposite in direction is exerted on the first object (National Research Council 1996, 180).

Materials

- Varying-sized ball bearings (purchased at a hardware store for about $1.50 each)

- 4 ft. piece of 3 in. diameter PVC pipe (purchased at a hardware store for about $4.00)
- Graph paper
- Scientific calculators
- Protractors
- Stop watches
- Scale

Engage

I begin by dropping a marble and asking my students why the marble fell to the floor. "The force of gravity made it drop," says a student. I follow up, "Okay, a force made it drop. What is a force?" Initially, my students respond, "It's a pull." Then, I pick up my pen from my desk and throw it across the room. One student immediately says, "It's also a push." I say, "A force is a push or a pull."

I then ask, "When I threw the pen across the room, at what point was my force no longer being applied to the pen?" Several students respond, "When it hit the wall your force was no longer on the pen." My students incorrectly believe that my force on the pen traveled with the object. To address this misconception I ask, "What is the result of a force on an object?" A student explains, "A force causes an object to move faster, or in the case of the wall, causes the object to slow down and change direction." Going back to my initial question about the force traveling with the pen, I inquire, "After I threw the pen, but before the pen hit the wall, could I have stopped the pen, changed the direction of the pen, or made the pen go faster?" At this point, students realize that my force on the pen ended when the pen was released from my hand.

Next, I pose the question, "Does a force always result in motion or a change in direction?" One student responds, "No." I went over and picked up the pen and placed it on a table. "Which forces are acting on the pen now?" A student replies, "The table and gravity." I ask, "Why isn't the pen moving if forces are acting on it?" A student replies, "Because the forces on the pen are equal." I then introduce gravitational and normal forces (Figure 7.1).

Explore

To get my students thinking about force and motion, I begin with a very simple demonstration. I ask my students to develop a procedure to determine how the angle of a tube affects the time it takes for the ball to travel down the tube. I tell my students that they have to incorporate ways to limit error into their procedure. This consists of doing each trial several times and taking an average.

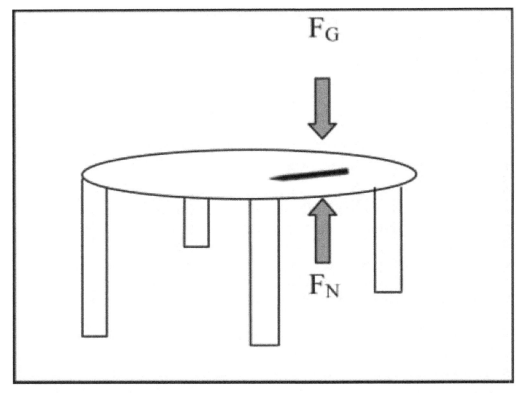

Figure 7.1 A motionless pen where FG = FN

Another way to limit error is to have students place their feet at the end of the tube so they can feel the ball hit instead of trying to "eye" when the ball hits the bottom (Figure 7.2). I also explain that they need to identify which variable they are changing, which variable they are measuring, and which variables need to be controlled. In this experiment, the angle of the tube is changing as well as the time it takes for the ball to travel down. The variable that must be controlled is the size and shape of the ball traveling down the tube on account of different shapes having different inertial moments (not because of different masses). Balls of different masses will travel down the tube at the same time. However, since students are conceptualizing force, and force is directly affected by mass, it is just as easy to use the same ball from one trial to the next. Students perform the procedure and record their results (Figure 7.3).

Explain

After my students have transformed their data into a graph, I ask for their observations. As expected, the trials when the tube was at a lower angle took longer

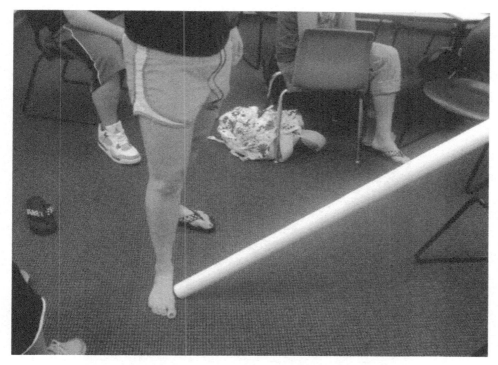

Figure 7.2 A student uses her foot to feel the ball hit her foot instead of having students try to "eye" when the ball will get to the end of the tube. After the ball hits her foot, she stops the timer. Although her reaction time is included in the overall time, it is relatively consistent (and much more consistent than "eyeing")

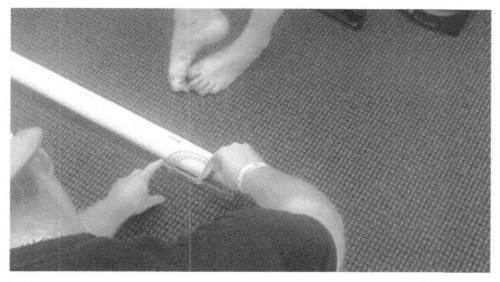

Figure 7.3 A student measures the angle of the tube prior to calculating the time it takes for the ball bearing to travel down it

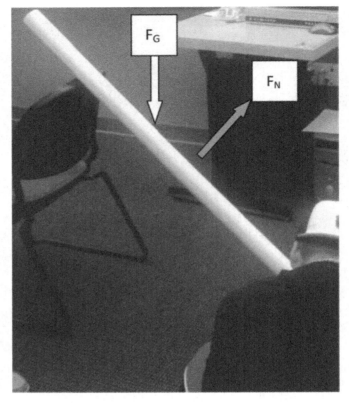

Figure 7.4 The force of gravity on the ball traveling through the tube is always in the y direction. The normal force of the tube acting on the ball is perpendicular to the tube

for the ball to reach the end. Instead of asking students to explain their observations, I ask them to identify the forces acting on the ball as it travels down the tube. "Gravity" is quickly announced. I refer back to the situation with the pen on the table. "When the pen was sitting on the table, was gravity the only force acting on it?" Students reflect and then explain that the tube was also acting on the ball. The normal force acts perpendicular (at a 90° angle) to the tube (Figure 7.4). Next, I pass out a piece of graph paper, which can be found free online at www. incompetech.com. On the graph paper, students see two ramps, Ramp A and Ramp B (Figure 7.5). Ramp A is at a steeper angle than Ramp B. For both situations (A and B), I ask students to draw the direction of the normal force, the force of gravity, the motion of the balls, and which ball will reach the bottom first if they are released at the same time and the hypotenuses of both ramps are equal.

Once students have had time to do this, I ask for their ideas about which ball will reach the bottom first and why. At this point, students understand that the ball going down the smaller ramp will take a greater amount of time to reach the

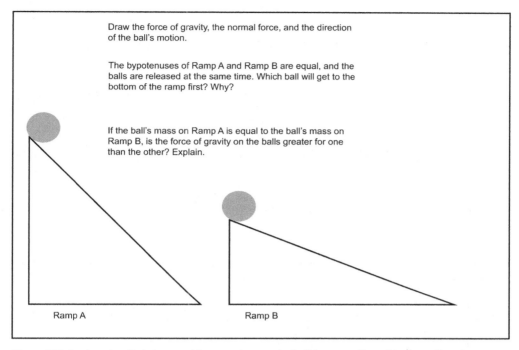

Draw the force of gravity, the normal force, and the direction of the ball's motion.

The bypotenuses of Ramp A and Ramp B are equal, and the balls are released at the same time. Which ball will get to the bottom of the ramp first? Why?

If the ball's mass on Ramp A is equal to the ball's mass on Ramp B, is the force of gravity on the balls greater for one than the other? Explain.

Ramp A

Ramp B

Figure 7.5 In situation B, the normal force acting on the ball is more aligned in the y direction

bottom. Students explain that the normal force becomes increasingly opposite to the direction of the force of gravity as the angle of the ramp decreases. I follow up by asking, "If I were to increase the angle from situation A to situation B, would the force of gravity on the ball change?" At this point, students understand that it is not the force of gravity that is changing; rather, it is the normal force (Figure 7.5).

Elaborate

At this point, I have not yet presented any mathematical calculations to my class. In the elaboration phase, I move from the conceptual idea of how the normal force changes, not the force of gravity, to a calculated explanation. This may require a refresher on the definitions of sin, cos, and tan theta (Figure 7.6). It may also require teaching students about vectors and that force is a vector. In a typical sequence, these concepts would have already been taught. I tell my students to remember the basics and take the calculations one step at a time. To keep things simple, the problem is identical to the inquiry they performed. In the problem (Figure 7.7), students calculate the force of a ball bearing traveling down two different ramps (Ramp A and Ramp B). For both ramps, the ball bearing travels 1.64 m (the hypotenuse of the triangle). I like

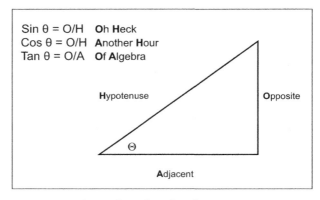

Figure 7.6 An acronym to remember sinθ, cosθ, and tanθ

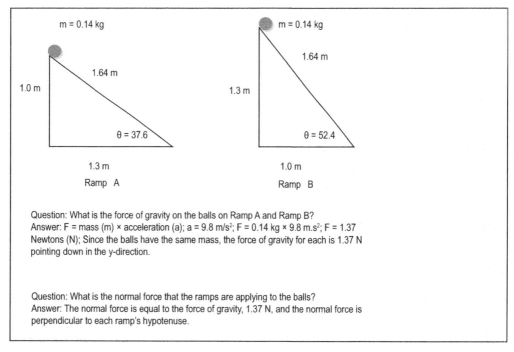

Figure 7.7 The force of gravity on the balls for Ramp A and Ramp B is equal. The normal force the ramp applies to the ball is the same magnitude, but in different directions

having the two situations side by side so that students—while they are cal-culating force, acceleration, and time for the ball bearings to travel down the ramps—can see how one variable (theta) specifically affects the calculations for finding the length of time it takes the ball to travel.

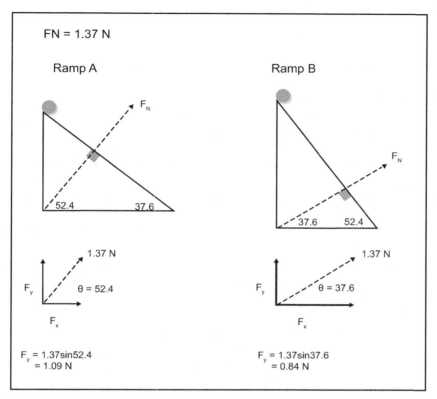

Figure 7.8 Focusing only on the normal force and dividing the normal force into x and y vector components

I show students step by step how to calculate the normal force in the *x* direction and the *y* direction (Figure 7.8), how to calculate the total force in the *x* and *y* direction (Figure 7.9), and how to calculate the force of the ball traveling down the ramp using the Pythagorean theorem (Figure 7.10). Once the force of the ball bearing down the ramp is known, the ball's acceleration (Figure 7.11) and the time it takes for the ball to travel down the ramp can be calculated (Figure 7.12).

The calculations are meaningful to my students because they had an opportunity to explore the science content. When I write the letters "Fa" and "Fn" and begin explaining the content in terms of vectors, *x* and *y* coordinates, and total force in the direction of the ball's motion, my students get it. They can connect the definitions to their experiences in the exploration phase. Although providing students with an exploration would seem to take more time, on the contrary it actually saves time. I do not have to teach and re-teach the same content. By encouraging students to actively explore the concept hands-on and getting students to conceptualize prior to calculating, I create opportunities for meaningful learning experiences.

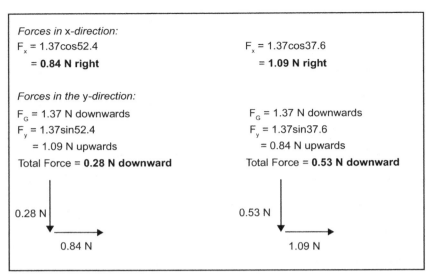

Figure 7.9 Calculation of the total forces in the x and y directions for Ramp A (left) and Ramp B (right)

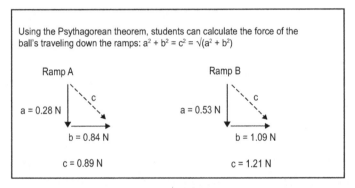

Figure 7.10 Students can calculate the force of the ball rolling down the ramp if they know the total force in the x and y directions

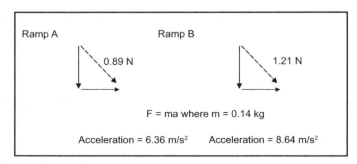

Figure 7.11 The acceleration can be calculated if the force of the ball rolling down the ramp is known

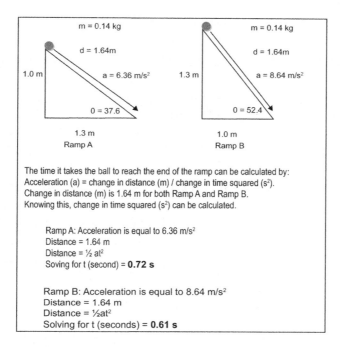

The time it takes the ball to reach the end of the ramp can be calculated by:
Acceleration (a) = change in distance (m) / change in time squared (s^2).
Change in distance (m) is 1.64 m for both Ramp A and Ramp B.
Knowing this, change in time squared (s^2) can be calculated.

Ramp A: Acceleration is equal to 6.36 m/s^2
Distance = 1.64 m
Distance = ½ at^2
Soving for t (second) = **0.72 s**

Ramp B: Acceleration is equal to 8.64 m/s^2
Distance = 1.64 m
Distance = ½at^2
Solving for t (seconds) = **0.61 s**

Figure 7.12 The length of time it takes the ball to reach the bottom of the ramp

Evaluate

I evaluate my students in two ways. I first evaluate their abilities to do inquiry. I ask my students to determine if the mass of a ball bearing has any effect on the time it takes for the ball bearing to reach the bottom of the tube. Students develop a procedure whereby they use various sizes of ball bearings (Figure 7.13), and as before, they calculate with a stopwatch the length of time for the ball bearings to travel down the tube. To ensure that

Figure 7.13 For the evaluation phase, I have students use different-sized ball bearings to determine how the size of the ball bearing affects the amount of time it takes for the ball to travel down the tube

the angle of the tube does not affect the trials, the tube is kept at a constant angle from one trial to the next. Students proceed to do several trials for each size of ball bearing to ensure reliability. Second, I evaluate my students' ability to calculate force, acceleration, and time. In this regard, I ask students to determine the amount of time it takes two separate ball bearings of varying masses to travel down the same incline plane (Figure 7.14). The incline plane is 1 m high at a 45° angle. The first ball bearing is 10 g and the second ball bearing is 40 g. If students do the problem correctly (Figure 7.14), both balls reach the end of the ramp in the same amount of time, 0.61 s. In this evaluation, students are not only showing me that they can perform a simple procedure and calculate a hypothetical problem, but also, they are learning something new that I did not have to teach directly to them. This new concept is that the mass of the ball bearing has no effect on the amount of time it takes the ball bearing to reach the bottom of the tube.

Conclusion

This lesson is an effective approach to teaching forces relating to the movement of an object rolling down an inclined plane. It creates an opportunity for students to understand the concept because it emphasizes a conceptual approach, providing students with hands-on experiences and guiding student inquiry prior to "solving the problem." When students understand the concept, they not only can solve the problem; they can transfer what they know to unique situations.

This lesson will require approximately two to two-and-a-half 50-minute class periods. However, the lesson will take longer if your students get some unexpected results. For example, it would be unexpected if the angle of the tube has no effect on the time it takes the ball to travel down the tube. Also, it would be unexpected if in the evaluation phase students observe the smaller balls reaching the bottom of the tube faster than the larger balls. Reasons why students may get unexpected results are (1) the PVC pipe is not long enough to obtain a reasonably accurate time for the ball to travel; (2) students are not doing several trials and taking an average; (3) the balls used to travel down the tube are made of different materials resulting in different coefficients of friction; (4) some of the balls are solid and others are hollow (change in the inertial moment); and (5) some of the balls are sliding rather than rolling (this would be unusual).

Extension

An extension option to this lesson can be found on C. K. Ng's PHY-is-Phun website. The website contains a simulation whereby students can change

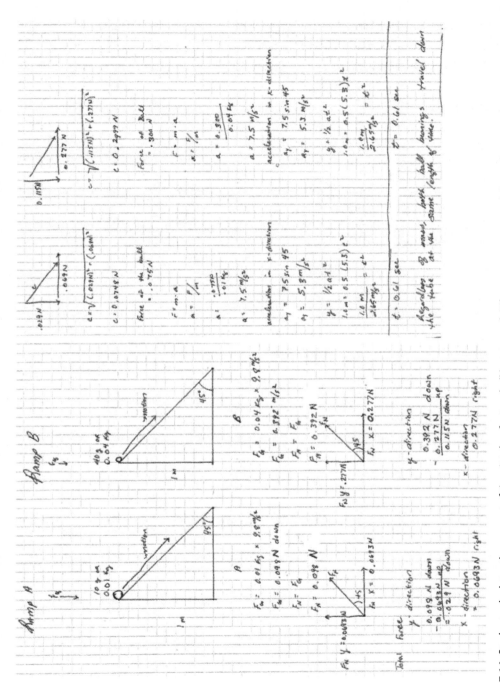

Figure 7.14 Student calculates the amount of time it takes for two ball bearings of different masses to travel down the same 45° incline plane

several variables (such as theta and the coefficient of friction) and see how those changes affect the time it takes an object to slide down a ramp (www. phy.hk/wiki/englishhtm/Incline.htm). Because this activity introduces the concept of friction, it lends teachers the opportunity for further discussion on the topic.

References

Hewitt, P.G. 2005. *Conceptual physics*, 10th ed. Reading, MA: Addison-Wesley.

National Research Council. 1996. *National science education standards.* Washington, DC: Academy Press.

National Research Council. 2005. *How students learn: Science in the classroom.* Washington, DC: National Academies Press.

Ng, C. K. 2009. *PHY-is-Phun: Learn physics using Java.* www.phy.hk/wiki/englishhtm/ Incline.htm (accessed February 23, 2011).

8

Model Lesson 6: Students' Investigations in Temperature and Pressure

Patrick Brown, James Concannon, Bernhard Hansert, Ron Frederick, and Glen Frerichs

About This Lesson

Why does a balloon deflate when it is left in a cold car; or why does someone have to pump up his or her bike tires in the spring after leaving them in the garage all winter? To answer these questions, students must understand the relationships among temperature, pressure, and volume of a gas. The purpose of the Predict, Share, Observe, and Explain (PSOE) activity is for students to connect the relationships among temperature, pressure, and the amount of gas. Students have many everyday experiences with these relationships, such as pumping up a basketball, but may have never thought of air in terms of pressure and temperature.

Why does a balloon seem to shrink when it is left in a very cold car, or why does a bike tire deflate after resting on a cold garage floor? To answer these driving questions, students must understand the relationships among temperature, pressure, volume, and number of particles in a closed system. However, studies show that students have difficulty understanding the underlying particle ideas about gases (Driver et al. 1994). In spite of the difficulties students face in learning about gases on a molecular level, they develop deeper conceptual understanding through exploration before an explanation instructional sequence that includes Predict, Share, Observe, and Explain (PSOE). During the lesson, it is essential that the teacher facilitates students' collection and interpretation of data and uses representative models to describe and predict "changes in particle motion, temperature, and state of a pure substance when thermal energy is added or removed. (MS-PS1–4)" (NGSS Lead States 2013). The sequence of activities helps students develop long-lasting understanding of the details of the relationship between the number of gas molecules, thermal energy, and empirical data. What follows is a PSOE sequence for how physical science students can quickly grasp an accurate understanding among pressure, temperature, and volume relationships.

Lesson Requirements

Materials
The following materials are needed:

- Cooler
- Ice
- Water
- LCD or small alcohol thermometer
- Ping pong balls
- Bicycle pump
- Pressure gauge
- 2-liter bottle
- Washer
- Large, clear, plastic container
- Nut

Preparation
Preparation for this demonstration is a must. To create an airtight 2-liter bottle that can be connected to a bicycle pump, first drill a hole in the cap large enough for a bicycle valve stem to fit through. A valve stem can be found at a hardware store, or usually at any location that sells bicycles. Run the valve stem through the hole in the cap. Secure the stem in the bottle cap using a washer and a nut. The rubber sleeve on the valve stem should fit into the hole in the cap, creating an airtight fit; then, place a washer through

Figure 8.1 Valve stem

Figure 8.2 Valve stem secured in cap

the valve and screw a nut onto the stem to limit the movement of the valve (Figures 8.1 and 8.2).

Once the valve stem is set in the bottle cap, place a LCD thermometer strip inside the bottle. The LCD thermometer strip provides a means to investigate any possible temperature change. If an LCD thermometer strip is unavailable, simply insert a short alcohol thermometer (Figure 8.3). Ideally, the temperature reading should be a thick red line for students to easily read. (*Safety Note:* Teachers should try the demonstration on their own multiple times using

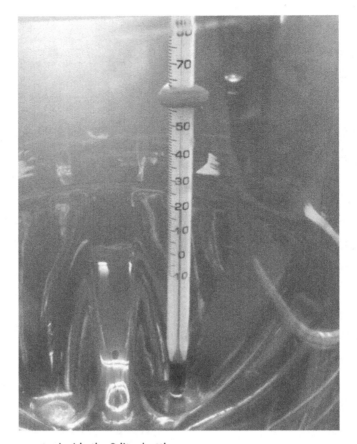

Figure 8.3 Thermometer inside the 2-liter bottle

protective eyewear before performing it for students. Students should always wear protective eyewear while doing this activity.)

Time
The time needed is three 50-minute class sessions.

Safety Considerations
Students should wear safety glasses when in proximity to a pressurized bottle. When pumping air into the bottle, 30 psi is the upper limit the bottle can withstand. Going above 30 psi is a safety hazard.

Targeted Grade Level
This activity is intended to help eighth and ninth grade students conceptualize and understand the relationships among temperature, pressure, and volume prior to introducing gas law formulas.

PSOE Demonstration

In this PSOE activity, students connect the relationships among temperature, pressure, and the amount of gas. Students have many everyday experiences with these relationships, such as pumping up a basketball, but may have never thought of air in terms of pressure and temperature. We begin the lesson by asking students to think about some of these everyday experiences, such as a balloon shrinking when it is left in a very cold car or a bike tire deflating after resting on a cold garage floor. Why does this happen?

Predict

Begin by showing the class a 2-liter bottle attached by a hose to a bicycle pump. Ask, "What do you predict will happen to this 2-liter bottle as air is pumped in it?" Have students write their ideas. Guide students into thinking about these questions:

> Will the size of the bottle change? Why or why not?
> Will the pressure in the bottle change? Why or why not?
> Will the temperature in the bottle change? Why or why not?

Students have varying ideas regarding the temperature of the bottle. Some students think that the bottle will get warmer, colder, or have no change. We found that approximately 47% of our classes thought that the temperature would increase, 11% thought the temperature would decrease, and 42% thought the temperature would not change.

Share

The purpose of letting students share ideas is to promote student-to-student conversations and allow students to justify their predictions. Students share their ideas for approximately 2 min. Based on their conversations, some students change their ideas or find that their ideas are supported by their peers. Many students use their firsthand experiences with 2-liter bottles (shaking a bottle and soda spewing everywhere) to try to convince their peers that their ideas are correct.

Observe

Next, cap the 2-liter bottle with the valve stem. Before pumping any air into the bottle, have a student measure the pressure using a tire gauge and the temperature inside the bottle. In the activity shown, the temperature of the bottle prior to pumping was 19°C (Figure 8.3) and the pressure was approximately 9 psi (Figure 8.4).

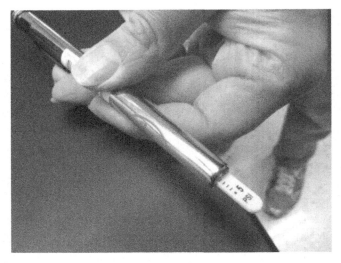

Figure 8.4 Measure the psi in the 2-liter bottle

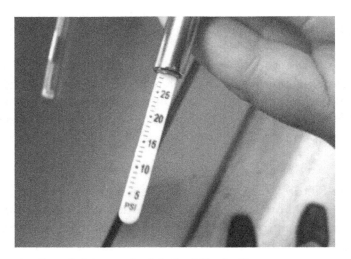

Figure 8.5 Measuring the psi after pumping into the 2-liter bottle

After recording the initial temperature and pressure (Figures 8.3 and 8.4), have a student (wearing safety goggles) begin pumping air into the bottle. It is helpful to have a bicycle pump that can measure the pressure inside the bottle as the student pumps; however, if this is unavailable, the pressure of the bottle should be measured with a tire gauge every five to six pumps to ensure the pressure does not exceed 30 psi (Figure 8.5). Once the gas pressure in the

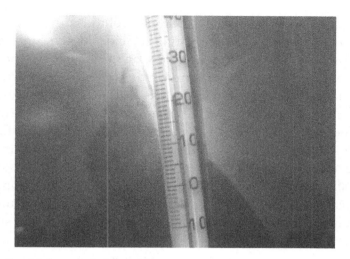

Figure 8.6 A noticeable increase in temperature

bottle reaches 30 psi, let the bottle stand for 5 to 10 min before having a student measure the temperature (Figure 8.6).

Though using an LCD thermometer would be much more convenient and accurate, students can see an increase in temperature using the alcohol thermometer. In this experiment, the temperature increased from 19°C to 21°C. Students quickly learned that as they increased the amount of air in the 2-liter bottle using the bicycle pump, the temperature inside the bottle increased.

Explain

The demonstration serves as a firsthand experience to engage students in a discussion about the kinetic molecular theory of gases. The discussion elaborates on students' experiences learning about gas molecules being in continuous, random motion, and that the speed of molecules, number of molecules, and molecular collisions all influence the energy (called "thermal energy" or "kinetic energy") in a system. Recap the results of the demonstration: (1) when pressure increased, temperature increased; (2) when pressure decreased, temperature decreased; (3) when temperature increased, pressure increased; and (4) when temperature decreased, pressure decreased. Based on the data collected, temperature and pressure are directly related to each other.

To help students make sense of difficult ideas, teachers can use ping pong balls to represent atomic and molecular movement of gas molecules and energy. The ping pong balls allow students to use model-based reasoning to critically think about the demonstration. First, place 12 ping pong balls in a transparent, 2-liter container. Close the container and have a student shake up the container so that the fast-moving particles collide with each other. Explain that when the ping pong balls are hitting the sides of the container, they are applying a force on the container. This force on the container is pressure. Pressure in our demonstration was measured in the unit of pounds per square inch, or psi. Pressure is directly related to the mass of the molecules and the velocity of the molecules. Temperature, on the other hand, is the measure of heat in the container. The amount of heat in a container is measured using a thermometer. Molecules collide into the thermometer, allowing it to measure temperature.

Next, have the student shake the container at different speeds. Shaking the container fast increases the velocity of the balls, similar to increasing the kinetic energy of gas molecules. The increased velocity of the balls resulted from an increase of energy into the system. The energy transfer into the system (heat) increased the velocity of the balls (kinetic energy) and therefore increased the temperature reading. Have the students imagine there was a thermometer in the container. When heat was added to the system, the ping pong balls would go faster and hit it more often, and, likewise, it would read a higher temperature. In addition to increasing the amount of heat, the pressure would increase as well because there are more collisions on the container walls. The molecules will be applying a greater force on the same amount of area. Next, open the container and add six more ping pong balls, for a total of 18 ping pong balls, and have the student shake up the container again. After the demonstration, ask students how increasing the number of ping pong balls (molecules) affected temperature and pressure, keeping the amount of heat into the container (the shaking) constant. Students understand that since there are more molecules in the container, there are more collisions in the container. Students understand that even if no heat is added or removed from the system (the intensity of the shaking), pressure will change if more molecules (ping pong balls) are added.

The ping pong ball activity provides visual support that pressure and thermal energy are directly related when the volume is constant. Students explain that as a result of the kinetic molecular theory, adding particles of gas to a container with a constant volume increases collisions, thereby increasing the pressure within the system.

Evaluation

To determine if students understand temperature and pressure and the relationship between temperature and pressure, pump air into the 2-liter bottle to approximately 30 psi. The bottle should still have the LCD or short alcohol thermometer inside it. Tell students to write down, in their own words, what happened to the motion of the gas molecules as more air was

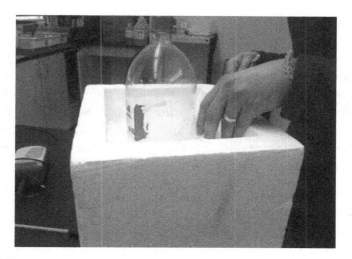

Figure 8.7 Packing the bottle in ice

pumped into the container. Students may also want to include diagrams with the written text. Have students present their explanation diagrammatically or using a combination of written text and other representations.

Next, take the temperature of the air in the 2-liter bottle at a pressure of 30 psi (for our demonstration, it was about 20°C). Share the temperature with the class. Next, place the bottle into an ice container (Figure 8.7). Have students write or draw what they think is happening with respect to temperature and pressure of the air inside the bottle. Why will the temperature go down? Students should rely on the data they collected from the class period before to infer relationships among pressure, number of air molecules, and temperature. What happens with respect to the pressure in the container? Also, have students write and/or draw a prediction of how and why the pressure in the container will change (if at all) after the air temperature inside the bottle drops. Collect students' papers for evaluation. Before the demonstration, have your students discuss in small groups their predictions. Once all students share their ideas, discuss students' predictions as a whole group.

Next, allow the bottle to sit on ice for at least 10 to 15 min. Read the new temperature and air pressure using the internal thermometer and a tire gauge, respectively. Note: The air temperature in the bottle will continue to drop over longer periods of time, and air temperature varies inside the bottle as the temperature decreases; however, the pressure in the bottle will drop noticeably within the first 5 to 10 min. The air temperature in the bottle will drop from 20°C to approximately 10°C relatively quickly, and the pressure drops from 30 psi to approximately 25 psi. Finally, submerge the bottle into a cooler of ice and leave overnight. At the beginning of class the next day, share the bottle's temperature and pressure reading.

The next day, after taking the last pressure and temperature readings, use a more application type evaluation. Ask students to use what they learned regarding the kinetic molecular theory of gases to consider whether the air pressure in a tire changes when driving. Students draw on their prior knowledge (e.g., rolling friction) and new ideas about the kinetic molecular theory of gases to explain that friction from the car tire and road increase the thermal energy of the air in the tire, causing the tire pressure to increase.

Conclusion

By using the PSOE demonstration model, teachers can incorporate some essential features of inquiry, all the while focusing on important *Next Generation Science Standards* (NGSS). For example, in this lesson the PSOE sequence helped students to gain deep conceptual understanding on the trans-dimensional nature of the NGSS (e.g., the connection among Science and Engineering Practices, Crosscutting Concepts, and Disciplinary Core Content). Students understood the kinetic molecular theory and the relationship between pressure and thermal energy when a container has a constant volume. As a result of the PSOE instructional sequence, students worked individually and collaboratively to construct ideas about the relationship among temperature, pressure, number of molecules, and the volume of a container.

The activities lend themselves well to elaborations where students use the gas law equations (Charles Law, Boyle's Law, and Combined Gas Laws) to complete computational problems that reinforce their experiences investigating the relationship among pressure, volume, and temperature in different situations. In many ways, the activities meet the learning goals (students were beginning to master the performance expectation outlined by the NGSS), and students made connections among the various activities to develop content understanding and critical reasoning skills to construct explanations from evidence.

References

Driver, R., A. Squires, P. Rushworth, and V. Wood-Robinson. 1994. *Making sense of secondary science: Research into children's ideas.* New York: Routledge.

NGSS Lead States. 2013. *Next Generation Science Standards: For states, by states.* Washington, DC: National Academies Press. www.nextgenscience.org/next-generation-science-standards (accessed September 1, 2014).

9

Model Lesson 7: 2-Liter Bottles and Botanical Gardens

Using Inquiry to Learn Ecology

*Patrick Brown, Patricia Friedrichsen,
and Lou Mongler*

About This Lesson

In the project presented in this article, high school students create and observe minieco-systems in an ecology unit designed around a 5E (engagement, exploration, explanation, elaboration, and evaluation) instructional model. Students choose a wide variety of organisms and use creativity to design miniecosystems.

Just outside the high school's door is a premier location to showcase nature's wonders. We use our immediate biological world and engage students in designing 2-liter-bottle ecosystems. The simple inquiry activities in this project increase student motivation while addressing several National Science Education Standards (National Research Council [NRC] 1996) for learning ecology (see Appendix). We describe this project using a 5E instructional model of inquiry learning that includes (a) engagement,

Table 9.1. Outline of Miniecosystem Project Based on the 5E Model

5E phase	Purpose	Learning activities	Time needed (min)
Engagement	Determine prior knowledge and engage students	Explore pond	50
		Design systems	2 × 50
Exploration	Collect data and conduct hands-on research to investigate ideas	Portraits of miniecosystems	25
		Construct a food web of miniecosystems	25
		Record observations in journals and through charts and graphs	15 daily
Explanation	Develop explanations on the basis of data collected in the exploration phase	Water cycle in miniecosystem	25
		Nitrogen cycle in miniecosystem	25
		Food chains and webs in miniecosystem	25
Elaboration	Apply knowledge to new situations	Write journal entries on pollution, human growth, and deforestation	50
		Everyday examples of the water, nitrogen, and carbon cycles	75
Evaluation	Assess understanding of unit concepts	Miniecosystem research symposium	50

(b) exploration, (c) explanation, (d) elaboration, and (e) evaluation (Bybee 2002). Table 9.1 presents an outline of the project, which includes several component activities.

Engagement

On the first day of the unit, we entice tenth grade biology students with a "life or death" challenge: to design a self-sustaining ecosystem that can survive for three weeks. We call our projects *miniecosystems*.

Because weather in the Midwestern United States is pleasant in the early fall, students eagerly go outside to explore nature through this hands-on investigation. The following criteria and guiding questions help students to design their miniecosystems and help us assess students' prior knowledge.

Miniecosystem Criteria:

- Must contain aquatic and terrestrial components
- Should include four organisms (no vertebrates)
- One organism must be a plant

Students usually want to know whether their entire grade for the project is dependent on their organisms staying alive for the duration of the three weeks. Many students ask, "How will our organisms get food and water if we seal our miniecosystems and do not put in additional foods?" Others want to know, "How will our insects get oxygen if the bottles are airtight?" Although technically the miniecosystems are not 100% airtight—a small amount of atmospheric gases and water move in and out—they serve as models to discuss many different types of self-sustaining systems and cycles in an ecosystem. We use the following guiding questions with students:

1. What will be the oxygen source in your miniecosystem?
2. What will be the carbon dioxide source in your miniecosystem?
3. How will plants and animals obtain water?
4. What will serve as food sources for the organisms in your miniecosystem?

We use one 50-minute class period to collect organisms, but, given student excitement, this is barely enough time. Adjacent to our school is a large pond, and we take a field trip to the pond to collect organisms. For many students, the challenge is so alluring that they search outside of school time with their friends for unusual specimens. If an outside source of organisms is not readily available, inexpensive materials such as elodea, snails, soils, and seeds can be purchased. We check students' designs to ensure that appropriate organisms are used; students are not allowed to use vertebrates such as small mammals or fish.

When it is time to build the miniecosystems, we share with students examples of models from the book *Bottle Biology* (Ingram 1993) and investigate real-life contained systems such as the Biosphere II Project. We encourage students to be creative when designing their miniecosystems, and, because each system must contain linked aquatic and terrestrial components, students have to draw on their knowledge of the water cycle, the oxygen-carbon dioxide cycle, and nutrient cycles before beginning.

Some students use two bottles attached one on top of the other. First, they create an aquatic environment with pond water in the bottom bottle and cut off the top of the bottle that has the spout. Next, they connect an inverted

Figure 9.1 A student creates a detailed drawing of her miniecosystem

2-liter bottle to serve as a terrestrial habitat on top. They use the bottle cap as a filter by poking small holes to allow for the transfer of water but not terrestrial components. These types of systems are similar to the ones laid out for students in *Bottle Biology* (Ingram 1993).

We have been pleasantly surprised that many students use creativity to fabricate complex bottle systems. For instance, we have had successful four-bottle miniecosystems that contain two terrestrial and two aquatic habitats. Students who create these types of miniecosystems connect the bottles so they lie flat and form the outline of a square (see Figure 9.1). They use intricate wicking devices to ensure that water from the aquatic environment does not spill into the terrestrial system. Making students design their own systems allows us to assess their prior knowledge about the interdependence of organisms.

To avoid students cutting themselves with sharp scissors while trying to cut through the plastic bottles, have them burn a small hole with a heated nail or probe (see Figure 9.2). Once this is done, they can easily cut through the 2-liter bottles. During this process, we reinforce good safety procedures by having them wear goggles, aprons, and gloves (see Figure 9.2).

The last step before our students seal their miniecosystems is to use soil-test kits to determine initial phosphorus, potash, and nitrogen levels in the soil. Students can compare these levels with those they measure when they take apart their miniecosystems at the end of the three weeks. This activity thus requires students to make quantitative measurements to investigate decomposition and the cycling of nutrients.

Figure 9.2 Student poster showing all components of the three-week miniecosystem project

Exploration

Students marvel at the miniecosystems they design, and line up at the door before school to check their ecosystems. After the initial construction stage, we give students 15 minutes of class time each day to make observations and ask questions about their miniecosystems. One way to encourage students' curiosity is to use a Video Flex camera. Video Flex is a high-resolution magnifying device that interfaces with a television. It can function as a microscope attachment, but we use it by itself to explore the miniecosystems. For many students, this is a nature-based reality television program: worms and pill bugs are active, rustling through the dead leaves and soil; spiders build webs; and grasshoppers jump from twig to twig. Like the outdoors, our miniecosystems do not allow anyone to edit out real-life events: students are touched by death and reconsider their initial design choices. The Video Flex camera allows us to use technology to increase students' appreciation for the detail found in nature, which reinforces the impact of humans on other living things (NRC 1996).

Explanation

After building and observing their miniecosystems, students are able to use evidence from the exploration activity to explain new concepts. We ask students to draw a scale image of their miniecosystem to illustrate the numerous cycles that occur there. For instance, students draw arrows showing that oxygen produced by plants cycles to the animal life, and carbon dioxide produced by animals cycles to plant life. Additionally, students infer that water accumulates on the inside of the 2-liter bottles because of evaporation from the aquatic habitat, and accumulated water condenses to form precipitation. Students answer the following guiding questions:

1. From where do plants get energy?
2. How is water cycled from the water habitat to the air?
3. How is water cycled from the terrestrial habitat to the air?
4. In what direction does energy flow in your miniecosystem?
5. What happens to animals after they die?
6. How are all the organisms in a habitat interconnected?

The miniecosystem mimics a real, natural system while allowing students to control variables such as animal and plant life. The guiding questions thus encourage students to use their observations of the miniecosystem to explain how biological systems in general work (NRC 1996). We find that students have the most difficulty explaining food webs. Library research on their organisms can help students make sense of their observations and to construct food webs of their ecosystems.

Students also may have difficulty understanding carrying capacity and logistic growth graphs. We discuss with students that resources and available habitat are a limiting factor to how large a population can grow. Students work in small groups to brainstorm factors that might limit a population in their miniecosystem from growing infinitely large. Then, we reinforce students' ideas with the carrying capacity growth curve. To explain logistic growth to students, we ask them to draw a graph of a population without any limiting factors.

Elaboration

Because many students are excited about their miniecosystems, we capture this enthusiasm to extend learning to real ecosystems. Students learn about food webs, the carbon-oxygen cycle, the nitrogen cycle, and the water cycle in more complex systems. Some students will discover through their library research that not all food webs display predator-prey interactions. This is a good opportunity for interactive discussions and writing activities on symbiotic relationships such as commensalism, parasitism, and mutualism.

During this unit, students ask many questions about human population growth, global warming, and deforestation. We ask them to address these topics in journal reflections about ecology-related current event articles that they read in the media. We find that letting students choose articles on these topics strengthens their understanding of the relationship between science and society (NRC 1996).

Evaluation

As teachers, we enjoy the activities in the evaluation phase because we use student presentations as an alternative form of assessment. Students summarize their understanding of ecology through a poster presentation at our miniecosystem research symposium (see Figure 9.2). During our symposium, students use evidence from their miniecosystems to demonstrate their content knowledge and question each other about the interactions that occurred in their systems. Students like to learn about their classmates' miniecosystems and whether their selected organisms survived the three-week project. The presentations are an excellent chance for students to communicate their understandings of ecology gained through the use of scientific inquiry of their miniecosystems (NRC 1996).

To give our students a wider audience, we ask them to display their posters in our teachers' lounge. This aspect of the project provides an opportunity for other faculty members to see how three weeks of inquiry learning comes together in a formal, culminating project. We find that many faculty members want to know more about using a 5E lesson to motivate students and that many students talk about the project in other classes. Students enjoy an extra sense of reward from the feedback they receive from other teachers.

Conclusion

Using a 5E instructional model helps us design inquiry experiences for our students and motivates our students to learn ecology. Students like the fast-paced learning and get involved as they make daily observations, take measurements, answer guiding questions, and create models of their ecosystems. Using these hands-on guided activities at the beginning of the year gets students excited about science for subsequent units. In the future, we plan to extend the elaboration phase of this project to include school-wide conservation efforts and to raise community awareness of recycling and pollution issues.

Although our ecology unit is at the beginning of the school year, the spring would also be an excellent time to collect organisms and build miniecosystems. We have purposefully put it at the beginning of the year because the ecology unit leads directly into later units on photosynthesis and cellular respiration, and we then have real-life examples from which to draw upon to help students make sense of these complex topics. By letting students create their own miniecosystems, we allow them to develop a personal interest in the interactions present in their models. As a result, they learn fundamental ecology concepts through this inquiry approach.

References

Bybee, R., ed. 2002. *Learning science and the science of learning*. Arlington, VA: National Science Teachers Association Press.

Ingram, M. 1993. *Bottle biology*. Dubuque, IA: Kendall/Hunt.

National Research Council (NRC). 1996. *National science education standards*. Washington, DC: National Academy Press.

Appendix

National Science Education Standards Addressed by Miniecosystem Activity

K-12 Life Sciences Content Standards

- Plant cells contain chloroplasts, the site of photosynthesis. Plants and many microorganisms use solar energy to combine molecules of carbon dioxide and water into complex, energy-rich organic compounds and release oxygen to the environment. This process of photosynthesis provides a vital connection between the sun and the energy needs of living systems.
- Energy flows through ecosystems in one direction, from photosynthetic organisms to herbivores to carnivores and decomposers.
- Organisms both cooperate and compete in ecosystems. The interrelationships and interdependencies of these organisms may generate ecosystems that are stable for hundreds or thousands of years.
- Living organisms have the capacity to produce populations of infinite size, but environments and resources are finite. This fundamental tension has profound effects on the interactions between organisms.
- Human beings live within the world's ecosystems. Increasingly, humans modify ecosystems as a result of population growth, technology, and consumption. Human destruction of habitats through direct harvesting, pollution, atmospheric changes, and other factors is threatening current global stability, and if not addressed, ecosystems will be irreversibly affected.

Source. National Research Council. 1996. *National science education standards.* Washington, DC: National Academy Press

10

Model Lesson 8: Students Conceptualizing Transcription and Translation from a Cellular Perspective

James Concannon and Maegan Buzzetta

About This Lesson

It is difficult for students to conceptualize biochemical processes that are portrayed as two-dimensional figures in a textbook. Instead of relying on overheads, PowerPoint, or textbook figures, the authors have students imagine themselves actually being inside a cell. Students have a specific role in the cell: helping with the transcription and translation process.

One of the most difficult tasks for science teachers is to explain concepts that cannot be seen with the naked eye or even with a standard classroom microscope. Students develop some sort of personal and often incorrect conception of these processes. Good science teachers use models to explain the invisible. This allows the teacher to assess and guide students' ideas. For example, a teacher may have students use pipe cleaners and beads to create models of different types of macromolecules. In this lesson, students use simple models of various macromolecules. This lesson takes three full 50-minute class periods.

Purpose

On completion of the lesson, students should be able to describe the location and properties of DNA, the process of transcription and translation, and the role enzymes play in protein synthesis.

This activity addresses the following National Science Education Standard:

Life Science Content Standard C: As a result of their activities in grades 9–12, all students should develop understanding of the molecular basis of heredity. In all organisms, the instructions for specifying the characteristics of the organism are carried in DNA, a large polymer formed from subunits of four kinds (A, G, C, and T). The chemical and structural properties of DNA explain how the genetic information that underlies heredity is both encoded in genes and replicated. Each DNA molecule in a cell forms a single chromosome.
(National Research Council [NRC] 1996, 185)

Materials

- White pipe cleaners
- Beads
- Red markers
- Scissors
- Paper and pencil
- Green markers

Day 1

A driving question is a good way to get the lesson going (Krajcik, Czerniak, and Berger 2002). Ideally, driving questions should relate to students' misconceptions. Writing a driving question on the board and asking students to write a response is a good way to assess what they know, or possibly what they do not know. The driving question also gives students the opportunity to realize that they may not fully understand a concept. This is a good way to motivate students. Ask, "How does DNA encode for proteins?" and "What are proteins made of?" Give students about 3 min to write their responses on a piece of paper. Have the students turn in their responses.

Begin with a short video funded by the National Academy of Sciences and Pfizer found at www.youtube.com/watch?v=sf0YXnAFBs8 (Virginia

Commonwealth University n.d.). Before it starts, ask students to take notes about the history and properties of DNA while watching the video (this video is about 8 min long). While the video is playing, read through the students' responses to the driving question. Pick out some particularly interesting responses to discuss. Once the DNA video has finished, read the selected responses to the driving questions out loud. Ask the students, "Now that you have taken some notes and watched the video, what are your thoughts about these answers?" Take several students' answers before asking, "What did you find out about DNA from watching this video?" Write students' responses on the board. Before proceeding to the next activity, make sure that students know that DNA is helical and double-stranded and that the strands are held together by bases. These bases are Adenine (A), Thymine (T), Cytosine (C), and Guanine (G). Next, pass out the DNA strands showing the 3' to 5' direction, but without base pairs (Figure 10.1). Students should work in groups of two to four.

After passing out the short DNA strands, explain to your students that they are creating their very own gene. Ask them, "What is missing in this segment of DNA?" Students should explain that DNA has complementary base pairs and that the base pairs are missing from the DNA strand. In this activity, students will create their own base pair sequence on the 3' to 5' strand. To determine the base pairs, pass out dice and a dice code. This code explains that each number represents a base (Figure 10.2). For six-sided

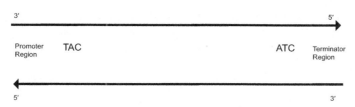

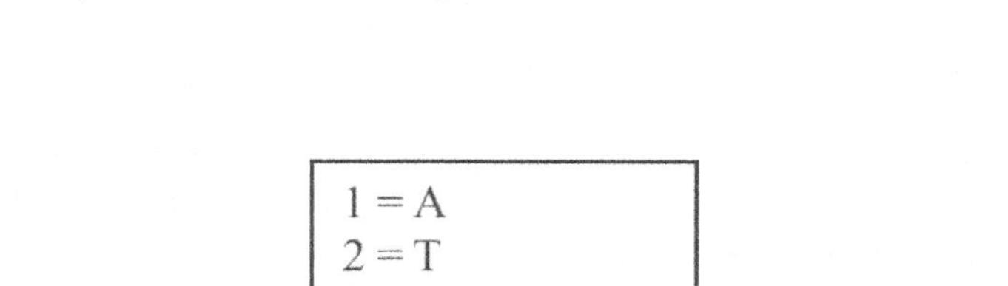

Figure 10.1 DNA sequence to use for transcription

$$1 = A$$
$$2 = T$$
$$3 = C$$
$$4 = G$$
5, 6 Repeat roll

Figure 10.2 Dice code

Figure 10.3 Example of possible student completion for 3′ to 5′ sequence

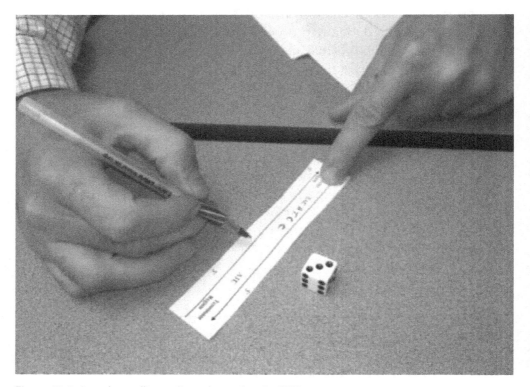

Figure 10.4 A student rolling a die to determine the DNA sequence

dice, have students repeat rolls for numbers 5 and 6. Have students roll the dice enough times to get 15 bases between TAC and ATC (Figures 10.3 and 10.4). After this, have students write down complementary base pairs for the 3′ to 5′ sequence on the 5′ to 3′ sequence (Figure 10.5). The teacher then describes the purpose of the promoter and terminator. Before the end of the class period, have the students write their names on the back of their DNA stands and hand them in.

To demonstrate the specifics of transcription, students make bracelets that represent mRNA. The mRNA bracelets are made from beads and white

Figure 10.5 Example of possible student completion for complementary base pairs

Figure 10.6 A mark Inside the loop shows where the mRNA begins

pipe cleaners. Begin this activity by explaining that the room is like a cell. Tell students that their desk is like a nucleus and that the DNA is inside the nucleus. Point out that the entire classroom is like a plant cell (with a cell wall present). The DNA is to stay on the desk because DNA transcription occurs in the nucleus. Students tie a loop at one end of the pipe cleaner and mark it with a green marker (Figure 10.6). The DNA that students created

Adenine, A: Purple
Uracil, U: Yellow
Cytosine, C: Blue
Guanine, G: Orange

Figure 10.7 Bead code for mRNA

Figure 10.8 String the beads according to the DNA sequence and the bead code

from rolling the dice serves as the instructions to make the bracelets. Four different colors of beads are needed. Each color bead represents a specific nucleotide (Figure 10.7). Students string a sequence of beads that correspond to the complementary bases of the 3′ to 5′ template (Figure 10.8). *Note:* The mRNA sequence is the same as the 5′ to 3′ DNA sequence (except for the thymine bases); however, remember that it is the 3′ to 5′ sequence that is used as a template for transcription. Students string each bracelet until they reach the terminator, and then they color the end of the pipe

Figure 10.9 Pipe cleaner with all the beads present

cleaner with a red marker (Figure 10.9). Students thread the end of the pipe cleaner through the loop and twist to form the bracelet (Figure 10.10). A possible question to ask students is, "Why is the mRNA and the 5′ to 3′ DNA strand so similar?"

Following this activity, explain to students that the 3′ to 5′ strand of DNA is used as a template for the production of mRNA. The process by which mRNA is produced is called transcription. The DNA is as it was before the transcription process. Because the students created the mRNA sequence, their involvement is synonymous at the cellular level with the enzyme mRNA polymerase. To reinforce this idea, have students explore the transcription process using a computer simulation called the Virtual Cell Animation Collection funded by the National Science Foundation and the U.S. Department of Education at http://vcell.ndsu.nodak.edu/animations/transcription/index.htm (Clark et al. 2008a). Pass out the questions found in Appendix A for students to answer while watching this video. The video is relatively short, so allow students to play it several times if necessary to answer the questions.

Figure 10.10 Completed bracelet

Day 2

On the second day, students do a simple activity to explore translation. For this activity, the teacher should have copies of the genetic code for each student. The genetic code does not contain bases; rather, it includes colors. The colors on the genetic code (see Figure 10.11) should match the colors used for mRNA. Make sure this is the case. If the teacher uses different colors for the mRNA, the colors of the genetic code should be changed respectively.

At the beginning of the class period, remind the students that the classroom represents a plant cell. They can imagine that their desk is the nucleus. (Remind students that, although there are several desks, a plant cell contains only one nucleus.) The day before, students worked at their desks to make bracelets that represent mRNA. These bracelets were made at students' desks because transcription occurs in the nucleus. At a separate location in the room (such as the lab table) are a stack of papers

		Second base				
		YELLOW (U)	BLUE (C)	PURPLE (A)	ORANGE (G)	**Third base**
First base	YELLOW (U)	Phenylalanine Phenylalanine Leucine Leucine	Serine Serine Serine Serine	Tyrosine Tyrosine *Stop* *Stop*	Cysteine Cysteine *Stop* Tryptophan	Yellow (U) Blue (C) Purple (A) Orange (G)
	BLUE (C)	Leucine Leucine Leucine Leucine	Proline Proline Proline Proline	Histidine Histidine Glutamine Glutamine	Arginine Arginine Arginine Arginine	Yellow (U) Blue (C) Purple (A) Orange (G)
	PURPLE (A)	Isoleucine Isoleucine isoleucine Methionine, *Start*	Threonine Threonine Threonine Threonine	Asparagine Asparagine Lysine Lysine	Serine Serine Arginine Arginine	Yellow (U) Blue (C) Purple (A) Orange (G)
	ORANGE (G)	Valine Valine Valine Valine	Alanine Alanine Alanine Alanine	Asparagine Asparagine Glutamic acid Glutamic acid	Glycine Glycine Glycine Glycine	Yellow (U) Blue (C) Purple (A) Orange (G)

Figure 10.11 The genetic code, using colors instead of bases for determining the amino acid sequence

containing the genetic code. Explain to the students that, although they are holding their mRNA bracelets at their desk, or nucleus, the next step requires the bracelet to move to the second location (the lab table) and that the lab table is like the rough endoplasmic reticulum. It is here where protein is produced, using the mRNA as a template. Students pick up a paper containing the genetic code. Working at the lab table (rough endoplasmic reticulum), students look at the genetic code and describe what they see. Students should explain that the genetic code shows three different bases. At this point, it is our experience that students do not know that the bases they see on the genetic code are used for determining the amino acid sequences. Explain that the bases on the genetic code correspond to the bases on the mRNA. Every three bases on the mRNA represent a codon, and the genetic code is a key we use to determine the amino acid sequence for each codon. Tell the students, "Looking at your mRNA sequence, starting with the green end and ending with the red end, determine the amino acid sequence. In this process, you are like a ribosome and tRNA. The ribosome calls for the correct tRNA and the tRNA brings the correct amino acid to form the protein."

Students should go through this activity with relative ease (see Figure 10.12). When done, students have a sequence of six amino acids. Ask the students to turn to their neighbors on either side and compare their amino acid sequences. Ask, "What are some things that you notice are the same and some things that

Figure 10.12 A student using the genetic code to determine the amino acid sequence

you notice are different between your amino acid sequence and your neighbor's amino acid sequence?" Students recognize that all amino acid sequences start with methionine, and each sequence is six amino acids long. Follow up with the question, "Why are everyone's sequences exactly six amino acids long?" A typical response is, "Because the mRNA sequence is 21 bases long and the last three bases do not code for an amino acid." Follow up students' conceptions of translation using the Virtual Cell Animation Collection found at http://vcell.ndsu.nodak.edu/animations/translation/movie-flash.htm (Clark et al. 2008b). Pass out the questions found in Appendix B for students to answer while watching this video.

Follow-Up

Review the transcription and translation process by going over the questions found in Appendix A and Appendix B. Next, ask the driving question that was originally posed to the students: "What is the end product of transcription and translation?" Students reply, "Protein." Ask, "What is protein made of?" Students should be able to answer, "Amino acids."

Ask the students, "Why does mRNA need to be produced? Why can't DNA be the template for translation?" After students think about it for a while, write down their ideas and talk about it, explaining that using an mRNA transcript allows the DNA to remain safe in the nucleus of the cell. DNA is not only present in transcription, but also in cell division. DNA is replicated before cell division so that other cells have exact DNA copies. Therefore, DNA must remain safe in one location. Finally, using a transcript of DNA allows for multiple transcripts to be made for simultaneous translation of protein.

Why do identical twins look identical? Why might people in the same family resemble each other? Some students will quickly respond, "Because identical twins have the exact DNA, and siblings could inherit common DNA." At this point, students may not know that enzymes are made from proteins and that different proteins produced from transcription and translation give rise to different enzymes. Students have the misconception that one gene determines one physical characteristic through the production of one protein. This is incorrect. Rather, multiple genes and environment determine a physical characteristic (Shaw et al. 2007). It is important for students to understand that one gene is responsible for the production of one protein, not one enzyme. In most cases, two or more proteins combine to create a functional enzyme. Explain to the class that enzymes produce molecules that give rise to a person's physical characteristics. Because identical twins have the same DNA, the proteins that come together to create enzymes are identical. Additionally, by chance, siblings may inherit a similar combination of genes that give rise to common physical characteristics. This explanation leads into future lessons about meiosis, genotype and phenotype, and dominant and recessive characteristics.

References

Clark, J.T., L. Daniels, P. Juell, P. McClean, D. Schwert, B. Saini-Eidukat, B. Slator, J. Terpstra, and A. White. 2008a. Transcription. Video. http://vcell.ndsu.nodak.edu/animations/transcription/index.htm (accessed November 29, 2017).

Clark, J.T., L. Daniels, P. Juell, P. McClean, D. Schwert, B. Saini-Eidukat, B. Slator, J. Terpstra, and A. White. 2008b. *Virtual cell animation collection.* Video. http://vcell.ndsu.nodak.edu/animations/translation/movie-flash.htm (accessed November 24, 2009).

Krajcik, J.S., C.M. Czerniak, and C.F. Berger. 2002. *Teaching science in elementary and middle school classrooms: A project based approach,* 2nd ed. New York: McGraw Hill.

National Research Council (NRC). 1996. *National science education standards.* Washington, DC: National Academy Press.

Shaw, K.R., K. Van Horne, H. Zhang, and J. Boughmann. 2007. Essay contest reveals misconceptions of high school students in genetics content. *Genetics, 178*(1), 157–168.

Virginia Commonwealth University. n.d. *The secret of life.* Video. www.youtube.com/watch?v=sf0YXnAFBs8 (accessed November 24, 2009).

Appendix A: Questions about Transcription Video

1. How does transcription begin?
2. What is the purpose of the transcription factors?
3. What enzyme is used for the transcription process?
4. What is the final result of the transcription process?

Appendix B: Questions about Translation Video

1. What is translation?
2. What are the three stages of translation?
3. What are the ribosome subunits responsible for in translation?
4. What is the purpose of the tRNA (transfer RNA) in translation?
5. How does the peptide grow?
6. What is a codon and anticodon?

11

Model Lesson 9: Are You Teaching Your Students about Stem Cells?

*James Concannon, Patrick Brown,
and Trisha Brandt*

About This Lesson

This is an interdisciplinary lesson designed for middle school students studying landforms and geological processes. Students create a two-dimensional topographic map from a three-dimensional landform that they create using clay. Students then use other groups' topographic maps to recreate landforms. Following this, students explore some basic ideas about how landforms take shape and how they can change over time. As students work through three distinct learning-cycle phases of concept exploration, introduction, and application, they use art, language arts, and mathematical skills to strengthen or form new science and social studies concepts.

Students have several misconceptions about stem cells, stem cell research, and cloning (Freyermuth et al. 2008). The topics of these misconceptions range from the location to the purposes of stem cells. Many students believe that adult stem cells come from adults and that embryonic stem cells come from children and infants. Students may also believe that stem cells can be implanted into a woman's uterus to produce a baby and that all stem cells come from placenta or embryonic cells (Freyermuth et al. 2008). This activity was designed to address such misconceptions about embryonic and adult stem cells while also addressing a science content standard for grades 9–12 (National Research Council [NRC] 1996).

National Science Education Standards

This activity addresses Life Science Content Standard C: As a result of their activities in grades 9–12, all students should develop an understanding of the cell

> Cells differentiate, and complex multicellular organisms are formed as a highly organized arrangement of differentiated cells. In the development of these multicellular organisms, the progeny from a single cell form an embryo in which the cells multiply and differentiate to form the many specialized cells, tissues, and organs that comprise the final organism.
>
> (NRC 1996, 185)

Materials

- Six Styrofoam balls, each with a diameter of 6.35 cm
- Scissors
- Scotch tape
- Two 100 g skeins of white yarn
- One 100 g skein of red yarn
- One 100 g skein of blue yarn
- One 100 g skein of green yarn
- One 100 g skein of yellow yarn
- Construction paper
- Markers

Teacher Preparation

Before class, the teacher should divide the six Styrofoam balls into two groups of three. One group represents adult stem cells. Follow these instructions to create the adult stem cells:

1. Wrap three Styrofoam balls in different lengths of red yarn. For example, the first Styrofoam ball could be wrapped with 10 m, the second with 16 m, and the third with 24 m (see Figure 11.1).

Figure 11.1 The three styrofoam balls representing adult stem cells should contain different lengths of red yarn

2. Tie white yarn to the red yarn.
3. Wrap each ball with approximately 20 m of white yarn. At this point, the balls should look nearly identical—all white.
4. All three adult stem cells will produce only one type of tissue. In this lesson, all adult stem cells produce cardiovascular tissue.

The remaining three Styrofoam balls represent embryonic stem cells. Each embryonic stem cell is made of a Styrofoam ball, a strand of white yarn, and a strand of colored yarn. The first ball should have white and green yarn, the second white and blue yarn, and the third white and yellow yarn. The different colors represent the type of tissue each embryonic stem cell will produce. Follow these instructions to create the embryonic stem cells:

1. For all three Styrofoam balls, begin by taping the colored yarn and wrapping it around the ball (see Figures 11.2 and 11.3). Wrap 30–40 m of the colored yarn around each ball (see Figure 11.4).
2. Tie white yarn to the colored yarn (see Figure 11.5).
3. Wrap 15 m of the white yarn over the colored yarn (see Figure 11.6). After all the yarn is bound, the embryonic stem cells will all look white (see Figure 11.7).

In addition to preparing the adult and embryonic stem cells, we also created a tissue chart (Table 11.1). This chart identifies the type of tissues represented by the colors of yarn. In this example, white and red yarn represents cardiovascular

Figure 11.2 Each embryonic stem cell will divide and differentiate into a different type of tissue
This stem cell will differentiate into renal tissue

Figure 11.3 This embryonic stem cell will divide and differentiate into hepatic tissue

tissue, white and green hepatic tissue, white and blue renal tissue, and white and yellow gastrointestinal tissue. The tissue chart could be presented as a PowerPoint slide or on a large piece of construction paper posted on the board.

The Activity

The purpose of this activity is for students to be able to explain differences between embryonic and adult stem cells. However, before beginning the activity it is necessary to assess students' misconceptions about adult and

Figure 11.4 Approximately 30–40 m of colored yarn should be wrapped around the styrofoam ball

Figure 11.5 White yarn is tied to the green yarn

embryonic stem cells. This can be accomplished by simply asking students to write down their ideas about how they think embryonic and adult stem cells are similar and different. Students can discuss their ideas about this question in groups.

After group discussion, students can explore their conceptions using the Styrofoam ball stem cells and the tissue chart. Show the class the six stem

Figure 11.6 White yarn is wrapped around the ball

Figure 11.7 The embryonic stem cell is complete

All three embryonic stem cells appear white, but each will differentiate into a different type of tissue

Table 11.1 Tissue Chart

Tissue Type	Color
Cardiovascular	Red/White
Gastrointestinal	Yellow/White
Hepatic	Green/White
Renal	Blue/White

cells, making sure the students understand that three are adult and the other three are embryonic. Separate the class into six groups; ideally, each group will consist of five to six students. Each group will get its own stem cell. The students need an open area to do this activity. Have them push desks out of the way, if possible, so each group can form a circle free of obstructions. Tell the students that they are going to create a tissue. Explain that a tissue is a group of cells that perform a common function. Before starting the activity, make sure that the students in each group know whether they have an adult or embryonic stem cell.

Give the stem cell to one student in each group. Have the student hold the end of the yarn in one hand and the ball of yarn in the other. Tell the class that the ball of yarn is one stem cell. Instruct the students who have the balls of yarn to keep holding the end of the yarn and throw the ball to another student in the same group. All groups do this activity simultaneously. After the first throw, stop the activity and explain that the line of yarn running from the first person to the second represents a new cell. Then ask the second person to hold onto the yarn and throw it to a third individual in the group. Explain to the class that this second line of yarn, running from the second to the third person, represents one of the daughter cells created from the cell division of the first line.

Allow each group to continue throwing the ball from person to person until no more yarn is left on the Styrofoam ball. Each new line represents one of the new daughter cells produced from a previous cell division, and each throw represents a cell division. At this point, a large spider web-like object has been formed, which represents a tissue. As the students hold up this tissue, have them look at the tissue chart to determine the type of tissue that was formed from the division of the adult stem cell. All three adult stem cells will produce cardiovascular tissue. The embryonic stem cells will produce different types of tissues. After the activity, have the class regroup to share their findings. Ask one member of each group to identify the type of stem cell they had, adult or embryonic, and what type of tissue was formed. For future reference, record on the board each group's type of stem cell and the type of tissue it produced.

Ask the class to explain the differences in the types of tissues that were formed by the adult stem cells and the embryonic stem cells. On the basis of the activity, students should note that the adult stem cells all produced the same tissue type. However, the embryonic stem cells produced several different tissue types. It is important to guide students into thinking about not only the tissue types but also the variety of tissues that each type of stem cell produced. After students understand that adult stem cells are limited in their ability to produce different kinds of tissue, the terms *specialized* and *unspecialized* can be introduced. (See the glossary in the Appendix for additional basic terms and definitions.) Embryonic stem cells are unspecialized, whereas adult stem cells are specialized. Why are adult stem cells specialized? In the next part of the activity, students will answer this question as they further explore the differences and similarities between embryonic and adult stem cells.

The day after the initial activity, have students work in groups of three to do research online about differences and similarities between embryonic and adult stem cells. A good source for this research can be found on the National Institutes of Health (NIH) web page (2006). Students should review the seven short chapters about stem cells and the highlighted words linked to the glossary on this website. Using this website, students will be introduced to key terms such as *pluripotent*, *multipotent*, and *totipotent*. As they do their research, students should record their findings in two columns: one of similarities and the other of differences.

After allowing 30 min for online research, ask students to write a short explanation about why they think adult stem cells are specialized. (This should be relatively easy to answer with information from the NIH website.) Adult stem cells are already specialized because they were obtained from specific tissues, whereas embryonic stem cells are obtained from the inner cell mass of a blastocyst. An adult stem cell can only become the kind of tissue from which it was originally obtained. In addition to asking students why adult stem cells are specialized, ask students to write an explanation about the relationship between types of stem cells, their specialization ability, and their medical potential. Using the NIH web page, students should be able to complete the task within one 50-min period.

Conclusion

Stem cells are a subject with which all people should be familiar. However, most students will enter the science classroom with many misconceptions regarding this topic. The teacher has the responsibility to address these misconceptions to help students become scientifically literate in today's society. Teachers may use inquiry-based activities such as this one to help address student misconceptions. This activity provides students with a scenario and questions intended to drive their thinking. Given sufficient time, appropriate materials, and targeted questions, students can address their misconceptions and develop a better understanding of stem cells.

References

Bioscience Network. StemCellResources.org: The science of education. www.stemcellresources.org/ (accessed November 28, 2017).

Freyermuth, S. K., M.A. Siegel, J. Concannon, and C. Clark. 2008. *Stem cells: Science and society.* Presentation at the annual meeting of Science Teachers of Missouri, Jefferson City, MO.

National Institutes of Health (NIH). 2006. Stem cell basics: What are the similarities and differences between embryonic and adult stem cells? In *Stem cell information.* Bethesda, MD: National Institutes of Health. https://stemcells.nih.gov/info/basics/2.htm (accessed November 28, 2017).

National Research Council (NRC). 1996. *National science education standards.* Washington, DC: National Academies Press.

Appendix

Glossary

Adult stem cell: A stem cell that is located in the tissues of a living organism. It is already specialized and is limited in its ability to differentiate. Adult stem cells are difficult to isolate because of their low percentage in tissue.

Blastocyst: A bundle of 70–100 cells developed from a zygote.

Clone: An organism with the same genotype as the donor parent.

Embryonic stem cell: A cell obtained from the inner cell mass of a blastocyst. An embryonic stem cell can differentiate into any type of tissue.

Reproductive cloning: A process leading to a live organism that is genetically identical to the parent.

Somatic cell: A diploid (non-sex) cell obtained from any tissue in the body.

Somatic cell nuclear transfer: The process of cloning. This process involves taking a somatic cell nucleus, containing all pairs of chromosomes, from a somatic cell and transferring it to an enucleated egg cell. The cell is then shocked to make it begin dividing.

Therapeutic cloning: Halting the cellular division at the blastula stage to harvest embryonic stem cells. Such stem cells can give rise to tissues identical to the parent's, thus decreasing the probability of tissue rejection.

12

Model Lesson 10: Transforming Osmosis

Labs to Address Standards for Inquiry

James Concannon and Patrick Brown

About This Lesson

A priority for all biology teachers must be for students to leave the classroom with a broad knowledge and understanding of science. Students need to be critical of science, analyze science, and relate new science knowledge to their daily lives. Unfortunately, many students are not reaching this goal. One strategy for making science laboratories more student-centered and to meet the National Science Education Standards for inquiry is to convert verification activities to inquiry-based investigations. The authors provide teachers with simple strategies to convert a confirmation-type osmosis laboratory into an inquiry investigation.

The overarching goal for biology teachers should be for students to leave the classroom with a broad knowledge and understanding of science. Students should be critical of science, analyze science, and relate new science knowledge to their daily lives (American Association for the Advancement of Science [AAAS] 1989; Bybee 2002). Unfortunately, many students do not achieve this goal. The National Science Education Standards (NRC 1996) emphasize that biology teachers need to address the concept of osmosis. Osmosis is a cellular response: the movement of water that maintains internal cellular stability while external conditions change (NRC 1996). Although students learn about osmosis, they leave high school with deeply held misconceptions about this phenomenon (Driver et al. 1994): they have difficulty understanding the flow of water across a semipermeable membrane and believe that there is no water flow when a solution is isotonic (Odom 1995). This is a pertinent issue in science education because "[a]n understanding of osmosis is key to understanding water intake by plants, water balance in aquatic creatures, turgor pressure in plants, and transport in living organisms" (Odom 1995, 409). The middle-level lab we present in this lesson provides students with conceptual understanding so they can build deeper knowledge before they reach high school.

Verification Labs

Why should teachers go beyond using verification laboratories in science classrooms? Verification labs focus on science terminology, concepts, and facts rather than students' prior experiences and knowledge. Verification labs lack meaningful and relevant driving questions that should focus the lesson and engage students in the material. Rather, the introduction is usually a paragraph explaining the phenomenon the students are going to observe. Following the introduction are "cookbook" procedures (i.e., step-by-step procedures that will help students arrive at a correct answer). The conclusion of a verification lab often asks students to write how they arrived at the single correct answer. Verification labs assume that if students "read it," and "do it," then they "learn it." In this way, students who do verification labs are denied the opportunity to think and reason for themselves—they are not allowed to ask and seek answers for their own questions. Last, verification labs do not ask students to reflect on and evaluate their new ideas in light of old understanding. One strategy for making science laboratories more student-centered, and to meet the National Science Education Standards for inquiry, is to convert verification activities to inquiry-based investigations.

Four Steps to Incorporating Inquiry

1. *Engage students in driving questions*
One entry point to an inquiry-based unit is developing a driving question. Meaningful driving questions situate learning in a relevant context while addressing science content (Krajcik, Czerniak, and Berger 2002). For example,

middle school teachers can begin their osmosis unit by asking the driving question "How can you keep vegetables crisp?" This is a good driving question for several reasons: it is open-ended, relevant to the topic (osmosis), and cannot be understood without knowledge of the content intended to be covered, and lends itself to researchable subquestions that students can examine on their own or in collaborative groups (Krajcik, Czerniak, and Berger 2002).

2. *Allow students to create a strategy to explore their predictions*

In the first 10 min of a class period, divide the students into groups of three. This is a good way to begin because it focuses the attention away from the teacher and onto the students. The student groups sit at a lab table that has a 500 mL beaker containing 250–350 mL of water saturated with sodium chloride (the most common household salt), carrots, celery, and peppers. Each freshly cut piece of carrot, celery, and pepper needs to be about 3 in long (this activity will work best if you soak the vegetables overnight in distilled water). At this point, most students should realize that the purpose of the activity is to see what happens to the vegetable when it is placed in the saltwater. Appropriately, ask students to write their responses to the question "What do you think will happen to the vegetables when you put them in saltwater?" This provides students with the opportunity to reflect on what they know. Some possible predictions students may have are that the vegetables will change color, the water will change color, the vegetable will change in texture, or there will be a change in mass for some of the vegetables. In the groups of three, the students should discuss their predictions and reasoning (if any).

This activity guides students in self-assessment and orchestrates discourse among students about scientific ideas (NRC 1996). While the students are discussing, collect the predictions and read them silently. While students continue discussing in their groups, write the top four predictions on the board. After 5 min of group discussion, direct students' attention to the four predictions on the board and identify the predictions individually. Next, tell the groups to determine four simple procedures that could determine if the four predictions are correct. Emphasize that the procedures should be structured so that their outcomes (i.e., the evidence collected) solely reflect the predictions. Ideally, the procedures will require the students to take measurements such as length and mass. For this reason, a ruler, graduated cylinder, piece of string, and balance should be made available.

3. *Provide materials and time required to perform the investigation*

Students should then begin their investigations. When students perform this activity, it is important that each wears safety goggles and washes his or her hands after putting them in saltwater (saltwater burns and dries out eyes).

This ensures a safe working environment (NRC 1996). At this time, teachers need to discuss the variable of time, or change in time ("How long are the vegetables going to be in the saltwater?"). A set time of 30 min for the vegetables to soak is sufficient. After the 30-min time period, students should collect evidence, paying attention to changes in the vegetables, and each group should discuss changes in the vegetables with respect to the top four predictions.

4. *Encourage students to reflect on their results to guide future explorations*
The students should have some unanswered questions after the initial investigation. These questions may include:

- Did the salt in the water cause the vegetable to become bendable?
- Does the length of time the vegetable sits in the saltwater determine how bendable the vegetable gets?
- Does the amount of water in the beaker determine how much the vegetable will change in mass?
- Does the initial volume of water or mass of the vegetable affect how bendable or crisp it becomes?

Ideally, each group has different questions to explore further; however, this is not always the case. The teacher needs to tell the groups to develop a procedure that allows them to answer one of their questions. It is important that the teacher walks to each group, asking the students what questions they have and checking that the group has developed a procedure that allows it to properly investigate the question. The question should be phrased so that it can be answered through a scientific investigation (NRC 1996). Direct students away from using observational words such as *hard* and *bendable*; instead, tell the students to use quantitative units that reflect how hard or how bendable the vegetable is.

After data collection, students need to transform their data into graphic representations, analyze the graphic representations, report their findings, and identify new questions that have arisen. The findings should be a description based on their observations (NRC 1996). Inevitably, each group will finish its experiment at a different time. One way I solve this dilemma, while incorporating an essential feature of inquiry, is to explain to the students that, after each group finishes its experiment, the groups will present their findings to the class with a PowerPoint presentation or three-paneled poster. The following sections must be included:

- Title
- Variables
- Units of measurement

- Procedure
- Data
- Graphs
- Findings
- Conclusion that answers the questions "Did you keep your vegetables crisp? Why?"

Follow-Up Activities

Teachers should provide students with the time to reflect, by writing in a journal, how their observations compare to their former predictions. Students can explain that water goes in or out of the vegetables; however, having students explain the phenomenon on a microscopic level is more difficult. Students can explore osmosis in red onion cells using a microscope, and free software is available online that provides an animation of osmosis at a microscopic level. Regardless of the strategy used to investigate osmosis at the microscopic level, allowing the students to explore osmosis in the vegetables makes the experience more meaningful and the knowledge long term.

The next step after a microscopic explanation of osmosis is to tie the phenomenon to a biologically oriented driving question. Because the teacher guided students from the beginning and engaged the whole class in a common exploration, students are able to take ownership of their learning and perform subsequent investigations in small groups. Groups can explore driving questions such as:

- Is gargling with saltwater beneficial if you have a sore throat?
- Why do you have to soak an eggplant in saltwater before cooking it?
- Why does ham have a salty taste?
- How do plants obtain carbon dioxide?
- The teacher can discuss how different kinds of questions require unique experimental designs and/or library research (NRC 1996). The lesson concludes with each group presenting its driving question, sources of information and/or experimental design, data, and a scientific claim based on the data.

Conclusion

Students arrive in the science classroom with many different ideas based on their knowledge and experiences. Often, students have deeply held misconceptions. To avoid propelling these misconceptions, it is critical for teachers to first identify their students' misconceptions and then design lessons so that students have the opportunity to collect data and make scientific claims based on their experiences. By engaging students in driving questions, allowing them to give priority to evidence to make scientific claims, and having them justify their claims in front of their peers, teachers can shift the orientation of investigations from being teacher-centered and passive to student-centered, active learning experiences (NRC 1996). We find this transformation results in higher student engagement, students overcoming their misconceptions, and the realization that everyday questions can be investigated in science class.

References

American Association for the Advancement of Science (AAAS). 1989. *Science for all Americans.* New York: Oxford University Press.

Bybee, R., ed. 2002. *Learning science and the science of learning.* Arlington, VA: National Science Teachers Association Press.

Driver, R., A. Squires, P. Rushworth, and V. Wood-Robinson. 1994. *Making sense of secondary science: Research into children's ideas.* New York: RoutledgeFalmer.

Krajcik, J.S., C.M. Czerniak, and C.F. Berger. 2002. *Teaching science in elementary and middle school classrooms: A project-based approach*, 2nd ed. New York: McGraw Hill.

National Research Council (NRC). 1996. *National science education standards.* Washington, DC: National Academy Press.

Odom, A. L. 1995. Secondary and college biology students' misconceptions about diffusion and osmosis. *American Biology Teacher, 57*(7), 409–415.

13

Lessons Learned

Instructional planning is a much more deliberate process than just identifying activities to use with students. The manner in which we sequence activities, how we have students explore their world, and the phenomenon we hone in on to focus instruction are important to ensure students learn the intended material. The process of incorporating the NGSS into practice can be achieved by using the design principles illustrated in the model lessons. The Introduction, Chapters 1 and 2, and the model lessons have been an attempt to start the conversation about what effective teaching and learning looks like in real-life settings. We have tried to encourage your professional growth by connecting you with some of the research in science education and model lessons that illustrate the research in practice. Here we share three brief lessons learned from designing and implementing the model lessons with students.

Lesson 1

Explore before explain instructional sequences tap into students' innate curiosity and cultivate their abilities to do and know science. We have worked with many teachers who are nervous about letting students explore science at the onset of an instructional unit. These teachers comment that students need "mini" lectures and prior instruction on content in order for students to be successful. Explaining content first is counterproductive for many reasons. First, many novice science learners have no framework in which to contextualize new knowledge. Second, learning content that is disconnected from firsthand experiences contributes to students viewing content and practices as being separate.

Students need to interact with materials and objects in the real world in order to have a framework for building science knowledge. This is because, as learners, all of us try to fit new experiences with what we already know. Once students have experience, they need chances to make evidenced-based claims and have authoritative explanations. After all, learning is more than just "hands-on;" students need "minds-on" opportunities to construct knowledge in supportive environments. It makes sense that we will most readily learn if we can find the proper fit between our firsthand experiences and science knowledge.

Lesson 2

Becoming an inquiry classroom is all about adopting a less-is-more approach to teaching. It is impossible and unnecessary to cover the amount of content found in most textbooks. Thus, inquiry-based approaches require teachers to focus on the most important topics that are beneficial for lifelong science understanding. Once teachers have focused learning objectives on desired content, students need opportunities to explore content at a deep level through direct experiences that use part or all of the essential features of inquiry and the appropriate variation. From engaging in inquiry activities, students develop robust ideas about the natural world and valid and reliable ways to generate knowledge. The rewards of using inquiry cannot be highlighted enough. Inquiry leads to higher achievement, improved attitudes, and greater abilities to think logically and reason about data and evidence and science knowledge. Thus, inquiry-based approaches are an important foundation for classroom instruction and cultivate the skills necessary to becoming more self-reliant learners.

Lesson 3

Teachers can motivate deep student learning by framing experiences around a real-life scenario or phenomenon. Phenomena provide for many learning activities and offer the chance for students to develop deep conceptual understanding. Most importantly, within a phenomena-based instructional approach, students have immediate opportunities to explore science rather than receiving knowledge directly from the teacher. If students view science knowledge as being completely unrelated to their lives, they are not likely to remember it or view it as important. Honing in on phenomena that are relevant to drive instruction helps students understand the intended content and make connections between supporting ideas. Phenomenon-based teaching helps students construct an extensive network of understanding. The result is that students can better transfer knowledge and skills between topics in the area of study and among other science disciplines. Students who engage in phenomenon-based activities have an understanding of how science relates to their everyday lives and the larger overarching concepts, as well as making connections to smaller, supporting ideas.

Conclusions

Changing our practice to address the NGSS may require us to look inward and consider why some approaches are more beneficial than others. The more we can help students make connections between their firsthand experiences with data and the claims they construct, the better they will know science and be able to use science in their lives. As you contemplate the model lessons and information presented in the introductory chapters, keep in mind that the twenty-first century demands innovative ways of problem-solving and critical thinking within a content area. When you have time in your classroom, take a moment to think about your next lessons. Questions should come to mind such as, "Do my students have firsthand experiences with science before teacher explanations?" "Do I use inquiry in my daily practices?" and "What phenomena drive my lessons?" This exercise will help you bridge the skills necessary to be successful in today's society with the research on how students learn science best. Right now, you are working towards developing more professional practice that will benefit students in the long term. We wish you the best of luck as you continue growing your repertoire of instructional practices that take students to higher levels of learning.